Sharks, Skates, and Rays

of the Gulf of Mexico

Sharks, Skates,

and Rays
of the Gulf of Mexico

A FIELD GUIDE

GLENN R. PARSONS

University Press of Mississippi
Jackson

Sea Grant
Mississippi-Alabama

This publication was supported by the National Sea Grant College Program of the U.S. Department of Commerce's National Oceanic and Atmospheric Administration under NOAA Grant # NA16RG2258, the Mississippi-Alabama Sea Grant Consortium, The University of Mississippi, and The University of Southern Mississippi award #USM-GR01166/ OMNIBUS-ED-17-PD. The views expressed herein do not necessarily reflect the views of any of those organizations. This is publication #MASGP-05-034 of the Mississippi Alabama Sea Grant Program.

This publication was supported by a grant to the author from the Mississippi Department of Marine Resources.

www.upress.state.ms.us

Designed by Todd Lape

The University Press of Mississippi is a member of the Association of American University Presses.

First edition 2006

∞

Library of Congress Cataloging-in-Publication Data

Parsons, Glenn R.
  Sharks, skates, and rays of the Gulf of Mexico : a field guide / Glenn R. Parsons.— 1st ed.
      p. cm.
  Includes index.
  ISBN 1-57806-827-4 (pbk. : alk. paper) 1. Sharks—Mexico, Gulf of—Identification. 2. Skates (Fishes)—Mexico, Gulf of—Identification. 3. Rays (Fishes)—Mexico, Gulf of—Identification. I. Title.
  QL638.9.P37 2006
  597.3'09163'64—dc22   2005020242

British Library Cataloging-in-Publication Data available

To the three most important people in my life:
my loving wife, Cheryl;
my beautiful daughter, Erin; and
the coolest guy I know, my son, T.J.

CONTENTS

ACKNOWLEDGMENTS

Many people contributed their time, their advice, and their photographs to this book. Dr. Paul Lago and Dr. Greg Tolley reviewed drafts of the manuscript, and I am deeply grateful for their assistance and advice. Lew Bullock, Jim Bartlett, Dr. Ramone Bonfil, Jeff Brown, Dr. Steven Branstetter, Dr. John Carlson, Dr. Eric Hoffmayer, Joe Jewell, Lisa Jones, Kristi Killam, and Dennis Riecke contributed photos, specimens, advice, and time, and to them I am very grateful.

Many of my undergraduate and graduate students contributed their time in the field and laboratory, including Ginny Adams, Reid Adams, Dr. Rachel Beecham, Will Bet-Sayad, Brian Cage, Linda Lombardi-Carlson, Jill Frank, Angie Haggard, Jeff Horton, Dalma Martinovic, and Melissa Sandrene.

I thank the faculty and staff of the Gulf Coast Research Laboratory, in particular Jim Franks, Dr. Mark Petersen, Sarah LaCroix, and Dr. Chet Rakocinski. I also thank the Fish Biology class of the University of Mississippi for reviewing drafts of the identification keys and providing helpful comments. Dr. Henry Bart Jr. and the staff of the Tulane University Ichthyological collection provided fish specimens for photography as did Dr. Jeffrey Williams of the Smithsonian Institution's Division of Fishes. I thank the staff of the Dauphin Island Sea Lab Estuarium, in particular Brian Jones, who was kind enough to provide specimens and photos of sharks that we might not have otherwise obtained. I thank Dr. LaDon Swann, director of the Mississippi-Alabama Sea Grant Program, and Corky Perret, director of marine fisheries at the Mississippi Department of Marine Resources for their financial assistance. I am also grateful for financial assistance provided by Dr. Alice Clark, Vice Chancellor for Research at the University of Mississippi. I thank Lance Ripley and vice president and director of husbandry, John Hewitt, of the Audubon Aquarium of the Americas.

Many of the photographs for this book were taken during a shark survey supported by the National Marine Fisheries Service under MARFIN grant # NA77FF0548 awarded to the author. Thanks also goes to Mark Grace, Michael Hendon, and the staff of the National Marine Fisheries Service Pascagoula Laboratory for providing photographs. A portion of this book was produced during sabbatical leave granted to the author by the University of Mississippi.

Sharks, Skates, and Rays

of the Gulf of Mexico

Introduction

ABOUT THE BOOK

This book is written for the scientist, naturalist, commercial or recreational fisher, outdoor enthusiast, beach-goer, or anyone interested in identifying and/or learning about the sharks, skates, and rays of the Gulf of Mexico. Using this book requires no specialized knowledge of scientific terminology and practically anyone should be able to quickly identify and easily learn about the most common species that are found in the shallow Gulf waters. To expedite identification I have used line drawings and color photographs extensively. Likewise, I have developed an identification key that asks a series of questions to help you arrive at an accurate identification of any Gulf of Mexico shark, skate, or ray that you are likely to encounter. I have arranged each group by family and then included the most common species of that family. Having researched and examined many thousands of fish from the Gulf of Mexico over the past twenty-five years, I have included personal observations and anecdotes about various species that cannot be found in any other book. I have tried to include any interesting facts that I have gathered over the years, as well as practical information about shark fishing, shark attack, safe handling of sharks, shark conservation, stingray safety, and treating stingray wounds. For the most part I have included all species that occur out to about 100 meters (330 feet) depth but have also described some of the deepwater species. Although it is very unlikely that the outdoor enthusiast would ever

encounter one of the deep-water denizens, I have included them because they are unique and many readers will find them interesting.
I believe I have struck a reasonable balance between educating and
entertaining the reader. I sincerely hope you agree!

BASIC SHARK ANATOMY

Effective use of this book requires knowledge of some very basic
anatomy. When referring to a certain fin or other structure, it is helpful for the reader to have some idea about which structure is being
referred to. There are six different kinds of fins on a shark, the **first
and second dorsals**, the **caudal or tail**, the **anal**, the **pelvics** (there
are two), and the **pectorals** (also two). The fins of sharks are very
different from those of other kinds of fish. While typical fish have
rather thin, flexible fins that are very maneuverable, sharks have
thick, rigid fins that are much less movable. The caudal fin provides
all of the thrust for swimming. I can personally attest to the power
of the shark caudal fin, having been slapped many times by this fin
while handling sharks. Fortunately, these were fairly small sharks. A
large shark could likely break bones with a swipe of its caudal. While
thrust is provided by the caudal, maneuverability is provided principally by adjusting the pitch of the winglike pectoral fins and flexing
the body. Most sharks have **five pairs of gill slits,** but some deep-
water sharks have six and seven gill pairs. Sharks obtain oxygen by
bringing water in through the mouth and allowing it to flow over the
gills and out through the gill slits. The "nostrils" in sharks are called
nares. Sharks "smell" the water by detecting dissolved substances
when they pass through the nares. The **pelvic fins** in male sharks
are modified into clasper organs (see Reproduction section), used
during mating to transfer sperm to the female.

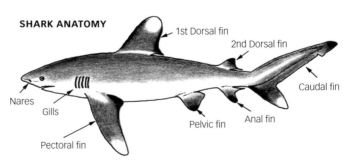

SHARK ANATOMY

1st Dorsal fin

2nd Dorsal fin

Caudal fin

Nares

Gills

Pelvic fin

Anal fin

Pectoral fin

BASIC SKATE AND RAY ANATOMY

While the basic plan of the skates and rays is similar to sharks, there are some important differences. The skates and rays are predominantly **benthic**; that is, they prefer life on the bottom of the oceans, and some spend time actually buried in the bottom with only their eyes protruding. Their body shape certainly belies their bottom-loving habit in that they are very flattened. The **pectoral fins** (see diagram) of the skates and rays are expanded and attached to the head. The flattened portion of the skate/ray body is referred to as the **disc,** and the disc width and shape are important in identifying different species. The length of the **snout** (measured from the eyes to the end of the snout) is also used for species identification. The gill openings of these fishes are located underneath the body. The skates differ in several respects from the rays. The most obvious difference between skates and rays is that skates do not possess a **stinging spine** on the tail whereas many rays do. The skates are harmless, although some have small **thorns**, particularly on the tail. The skates may also have a **caudal (tail) fin** and **first and second dorsal fins**; the rays have none of these. In some skates, the tail is known to contain an electric organ that is likely used in communication. The eyes of both skates and rays are located on top of the head. Just behind each eye is a fairly large opening. These openings are called **spiracles** and are actually used for breathing while the skate or the ray is

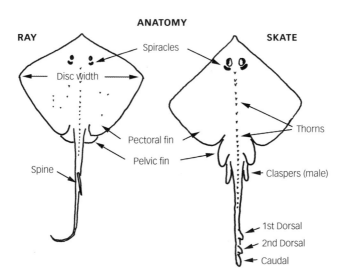

ANATOMY

RAY — Spiracles — SKATE

Disc Width

Thorns

Pectoral fin

Pelvic fin

Spine

Claspers (male)

1st Dorsal

2nd Dorsal

Caudal

lying on the bottom or is buried in the sediment. Typically, fish draw water into the mouth and then pump it out over the gills. However, because the mouth is located on the underside of skates and rays, it is not possible to draw water into the mouth without getting a mouthful of sand and mud. Therefore, these animals draw water in through the spiracles on top of the body and pump it out of the gills underneath. Skates and rays have **pelvic fins** and males, like sharks, have very prominent **claspers** that are used for introducing sperm into the female.

HOW TO USE THIS BOOK

I have attempted to make this book as user friendly as possible and have written it assuming no prior knowledge of shark, skate, or ray identification. If your goal is to identify some species, then the first decision to make is whether it is a shark, a skate, or a ray. This should be relatively simple, although there are a few fishes that can cause a problem. If the animal looks like a shark, it probably is a shark. If, however, it is very flattened, then it is likely a skate or a ray. There are some exceptions to the above. The sawfish are very sharklike in appearance but are, in actuality, elongate skates. Likewise, the guitarfish is an elongate skate that has sharklike characteristics. Finally, the angel shark is a very flattened shark that could be confused with a skate or a ray.

To identify organisms, biologists use a series of questions to help them arrive at a correct identification. Often two questions are posed. For example, if you wanted to identify different types of common fruits, the first two questions might be:

1. A. *Is the fruit round? Go to question #2.*
    B. *Is the fruit other than round? Go to question #3.*

This first set of questions would obviously separate oranges and plums from bananas and pears and so on. This list of questions is referred to as a "key," and because most resources use a set of two questions, they are sometimes referred to as a "dichotomous key." We will use keys throughout this book, but in some cases we will pose more than two questions at a time. Additionally, we have included many diagrams within the keys to help in your identification.

THE FIRST STEP IN IDENTIFICATION:
IS IT A SHARK, A SKATE, OR A RAY?

To begin identifying an unknown shark, skate, or ray that you have captured or observed, answer the questions and choose one of the diagrams below. This first very simple key will help to get you to the correct book section.

1. Is the body shape in general that of a shark? Go to page 49.

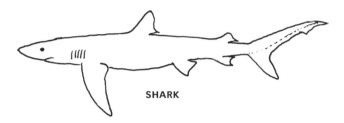

2. Is the body very flattened, with or without a long, slender tail? Go to page 119.

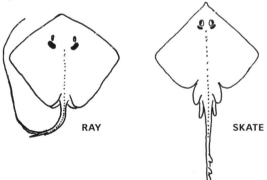

3. Is the body flattened but the pectoral fins *do not* attach to the head? Go to page 91.

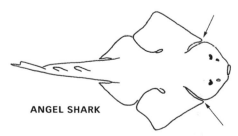

4. Is the body elongate but the pectoral fins *do* attach to the head? Go to page 126.

GUITARFISH

5. Is there a sawlike rostrum extending from the snout? Go to page 123.

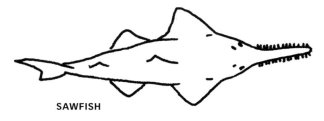

SAWFISH

The Gulf of Mexico

The largest gulf in North America, the Gulf of Mexico has an area of approximately 400,000 square miles and a coastline of about 2,500 miles. It extends from about 30° latitude in the north, 21° latitude in the south, 82° longitude in the east, and 97° longitude in the west. Water exchange between the Gulf of Mexico and the Atlantic Ocean is somewhat restricted because of the juxtaposition of the Yucatan Peninsula, Cuba, and Florida. Water flow in and out of the Gulf is by way of the relatively narrow Yucatan Strait and the Straits of Florida. The Loop Current is the major large-scale water-movement pattern with flow entering at the Yucatan Straits, looping up into the Gulf (the extent of this loop will vary with the seasons), and then exiting at the Florida Straits. The Gulf has a maximum depth of about 2,000 fathoms (12,000 feet), average water temperature

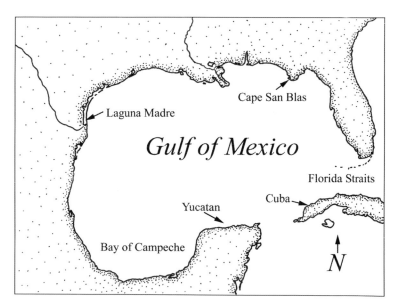

varies from about 12 to 30°C (53 to 86°F) in the north and about 28 to 30°C (73 to 86°F) in the south. This creates a warm-temperate environment for marine organisms in the north and a tropical environment in the south. The continental shelf, a region of relatively shallow water that extends out from the shore, is very broad in the Gulf, particularly along the Florida coast.

MAJOR FEATURES AND HABITATS

A number of major rivers empty into the northern Gulf but most notable of these is the Mississippi. In every respect, the Mississippi River is the largest river in the United States, and in terms of length, it is the fourth largest in the world. The Mississippi River discharges about 2 billion cubic meters of water and 300 thousand metric tons of sediment per day into the northern Gulf, having a significant effect upon salinity, sediments, and the marine habitats there. The input of sediments and rich organic materials make the northern Gulf extremely productive and support important crab, shrimp, and fish populations. In fact, the shrimp fishery in the northern Gulf of Mexico is the most valuable fishery in the entire United States with landings valued at almost $400 million in 2002. However, this nutrient input can be a mixed blessing. Nutrient input into the Gulf from agriculture, wastewaters, and the like has increased in recent years, and this has led to a seasonal depletion of the waters' oxygen content in a large area in the northern Gulf. This zone of depleted oxygen threatens the marine ecosystems and the important fisheries that exist there.

The Mobile Bay is a prominent feature in the north-central Gulf. This bay is characterized by reduced salinity and highly turbid waters. Sediment and nutrient input from the Dog, Fowl, Fish, Mobile, and Tensaw rivers contribute to great fish productivity. Although there are certainly disadvantages, the many gas rigs in the Mobile Bay area and offshore of the Alabama and Mississippi coasts attract fish and are good artificial habitat. I have captured many sharks around the gas rigs in this area. Because near shore water movements in the north-central Gulf are to the west, the river sediments that are delivered to this area are transported westward. This creates fine mud sediments and highly productive estuaries and marshes west of Mobile Bay. Another important feature of the north-central Gulf is the extensive system of barrier islands, including Dauphin, Petit Bois, Horn, Ship, and Cat islands. The waters behind these islands form the Mississippi Sound, a highly productive estuary. The waters of the Mississippi Sound can likewise be highly turbid, and salinity can be very low. The coast of Louisiana is dominated by the Mississippi Delta. The extensive marsh habitat in this area is dotted with literally hundreds of islands.

The Texas coast is also dominated by a system of barrier islands that extend almost uninterrupted from Matagordo Bay to the coast of Mexico. The Laguna Madre, an estuary formed by these islands, has exceedingly high salt content in the water because of restricted freshwater input and high rates of evaporation. From the U.S.-Mexico border to the Yucatan Peninsula are seen fewer barrier islands and consequently estuaries are less developed.

East of Mobile Bay, away from the influence of large rivers, sand and shell sediments are more prevalent. This lack of river water input is responsible for the snow-white beaches and clear-blue waters of Gulf Shores, Pensacola, Destin, and Panama City. While crystal-clear waters and sugar white beaches can certainly be a great place for bathers, in my experience it is not particularly good shark habitat, at least not during the day. The waters along open beach areas are typically not very productive so that there are not a lot of fish. Sharks prefer areas where there are lots of fish because, after all, that is what they eat. However, just east of Panama City around Cape San Blas is found some of the most exceptional shark habitat in the entire Gulf. I have observed incredible numbers of sharks in Crooked Island Sound just south of Tyndall Air Force Base and around Indian Pass.

The eastern Gulf of Mexico is formed entirely by the peninsula of Florida. Marshes and fertile oyster reefs are characteristic of the Apalachicola and Apalachee bays. The best-tasting mullet in the United States comes from these waters. South along the Gulf coast of Florida are found entangling mangroves, sponge communities and, in south Florida, coral reefs. The beaches of the central Florida Gulf coast are formed primarily from broken shell and are thus not as white as those in the northern Gulf. Tampa Bay, located in central Florida, Charlotte Harbor in the south, and Florida Bay, formed by the Florida Keys, are the three largest estuaries along this coast. Despite the extensive development around the shores of Tampa Bay it continues to support good fish populations. The Florida Middle Grounds are located between Apalachicola and Tampa Bay about 100 miles offshore and is about 350 square miles in area. Limestone ridges rise about 40 feet from the surrounding sand providing habitat for corals, sponges and, consequently, many species of fish. Charlotte Harbor, the Ten Thousand Islands area, and surrounding waters continue to be one of the most unspoiled estuarine habitats along the west coast of Florida. Florida Bay is a unique body of water because it is very clear and very salty, and it is considered a tropical environment.

While most are familiar with the shallow coastal areas of the Gulf, and this book will stress that environment in particular, by far the largest habitat for marine organisms is the deep bottom and open ocean waters. Unfortunately, we know relatively little about the sharks inhabiting the depths of the ocean. For example, as recently as 2002, we discovered that one of the rarest and strangest sharks in existence, the goblin shark, calls the Gulf of Mexico its home. As our knowledge of the deep waters of the Gulf of Mexico increases, there will no doubt be even more spectacular discoveries.

THE SHARKS, SKATES, AND RAYS OF THE GULF OF MEXICO

The Gulf of Mexico is blessed with a diversity and abundance of sharks, skates, and rays in its shallow coastal waters. However, it should be emphasized that far and away the most common sharks in the shallow waters of the Gulf of Mexico are small and relatively harmless. Of the approximately thirty-five or so species found in this area, only about five are considered dangerous, primarily because of their large size. The actual kinds of sharks you are most likely to encounter will differ depending upon where you are in the Gulf. In the southern Gulf waters and along the Gulf coast of Florida, the most common shallow-water shark is the bonnethead shark, a harmless species of hammerhead that grows to a maximum length of only about 1 meter (3.5 feet). Also very abundant are the blacktip, blacknose, and lemon sharks. In the northern Gulf waters the Atlantic sharpnose is the dominant species, with the blacktip and finetooth also very common. However, the bull shark is more common in and around the marshes of Louisiana. The bonnethead shark, while common in Florida waters, is not particularly common in the shallow waters of Alabama, Mississippi, and Louisiana.

In the cooler waters of the northern Gulf, the best months of the year for encountering sharks are from about April to November. During the fall, most sharks of this area migrate to warmer, offshore waters, and then return again in the spring. Some species, particularly the larger ones, engage in a more extensive migration and may move out of the Gulf of Mexico completely. For example, the migration of the sandbar shark between the Atlantic and the Gulf of Mexico is well documented. Likewise, it is known that the great white shark moves into the Gulf of Mexico during winter and early spring when water temperatures drop below about 22°C (71°F). We have observed white sharks captured from deep waters off of the central Florida Gulf coast in winter. In the southern Gulf, sharks may be captured year round.

Stingrays are common inhabitants of the Gulf, and they may be encountered almost any time of year but certainly are more common during spring, summer, and fall. I have captured many southern stingrays along the central and south Florida Gulf coasts. However, along the northern Gulf coast, the bluntnose and the Atlantic stingray are more common. Stingrays seem to be particularly common in very low salinity waters so they may be captured in estuaries, bays, near river mouths, and so on. The yellow stingray is commonly encountered on the grass flats of Florida Bay and along the Florida Keys. Skates are not common in inshore waters in the Gulf and prefer the more stable, higher salinity waters further offshore. However, fishing outside the barrier islands may produce either the clearnosed skate or the roundel skate, the most common species in the inshore Gulf of Mexico.

The Sharks

SHARKS: A PERSONAL VIEW

When I first decided to study sharks back in the late 1970s, my friends and family thought I was completely insane. My father, a U.S. Navy veteran who fought in the South Pacific, surely entertained thoughts of me being dismembered by these savage animals. Refer to any resource information on the sinking of the USS *Indianapolis* for a better understanding of a sailor's view of sharks. More than once I had to explain that the sharks that I studied were small (Atlantic sharpnose and bonnethead sharks) and that I did not have to be in the water with these animals (not entirely true, however). But the question of why I chose sharks as a study animal is a valid one. Although I never met a fish I didn't like and I cringe when friends and colleagues refer to me as a "shark biologist," sharks are singularly fascinating animals and they do occupy a special place in my psyche. The fact that they have survived and flourished for millions of years is a testament to their endurance. Who can possibly imagine the kinds of environmental challenges that sharks have faced and overcome in the hundreds of millions of years of their evolution? Most scientists today predict unprecedented environmental changes in the near term that have the potential to cause great harm to natural ecosystems. Along with these changes will undoubtedly come much human suffering. Perhaps the study of this very successful group of animals will provide insight into their resilience and answers to some of the problems we will assuredly face in the not-too-distant future.

SHARKS: FACT AND FICTION

The general public seems to have an insatiable appetite for anything shark. Unfortunately, many are fascinated for perhaps the wrong reasons. The attitude that people have toward these misunderstood animals ranges from unreasonable fear and disgust to rapt fascination. I was once a guest on a live radio talk show where a caller asked me if you would be immediately eaten if you fell out of the boat while fishing in the Gulf of Mexico. This person seemed to be suggesting that the sea was literally teaming with sharks ready to eat anything that came their way. I of course explained that the chance of even being bitten, much less devoured, would be exceedingly small. The more important question, however, is how could a rational person come to believe such a thing? Why are there so many myths and misconceptions surrounding sharks? Some of the misunderstanding has probably come about because of the sensationalistic news coverage that sharks seem to attract. It is true that we tend to fear what we do not understand, and sharks may be one of the most misunderstood animals in the oceans. The best way to overcome this fear is to educate ourselves, and I believe that addressing some of the facts and fiction regarding sharks will help.

FACT VS. FICTION

FICTION: Sharks are mindless eating machines ready to consume anything in their path.

FACT: Like all animals, sharks eat until they are full and will then stop eating.

FICTION: When sharks go into a feeding frenzy they will eat anything and have been known to even bite themselves.

FACT: It is true that sharks can be stimulated to engage in a kind of "mob feeding" behavior. However, this behavior appears to be an unusual (perhaps unnatural) behavior that is typically caused by the activities of humans. When unusually large quantities of fish remains or blood are placed in the water such as when trying to attract sharks, they may exhibit this behavior. However, most sharks would be unable to bite themselves during one of these frenzies because they are not flexible enough.

FICTION: When a shark tastes human flesh it will keep coming back for more.

FACT: There are very few sharks in the Gulf of Mexico that are large enough to eat humans, and these are typically found in the deeper waters well away from beaches and bathers. Additionally, the vast majority of sharks in the Gulf prefer food items that can be swallowed whole, such as whole fish, crabs, or lobsters.

FICTION: Sharks are scavengers and will eat most anything.

FACT: There are lots of different sharks with various dietary habits. In my research over the past twenty-five years I have examined the stomach content of countless numbers of sharks of various species. In all those years and all those stomachs I have found nothing more exotic than the occasional plastic bag (probably a discarded bait container) or small piece of driftwood (probably accidentally ingested while feeding). Sharks prefer fresh, preferably living, prey items. However, for one exceptional species see the description of the tiger shark.

FICTION: The most common shark in the shallow waters of the Gulf of Mexico is the sand shark.

FACT: The term *sand shark* is actually used to describe at least four different kinds of sharks, and in reality there is no shark in the Gulf whose common name is "sand" shark. Many coastal residents use this term for any small gray shark.

FICTION: There are lots of shark attacks that occur each year that never get reported in the news media because it hurts tourism.

FACT: In this day of twenty-four-hour news channels and unrestrained, often sensationalized media coverage, few shark attacks are likely to go unreported. You could almost say that there is a "feeding frenzy" of news reporters when an attack does occur.

FICTION: Scientists have discovered that sharks can actually stop swimming when for many years it was believed that they never stop. It appears that they can "sleep" under certain conditions.

FACT: This is probably the second most common question that is asked of me. The problem with this is that it is only half true. There are many kinds of sharks that can stop swimming and spend unlimited time resting on the bottom. There are also other kinds of sharks that can never stop swimming. If they stop, they will suffocate because they have to move water over their gills to get oxygen. Their forward motion is the same as you or I breathing.

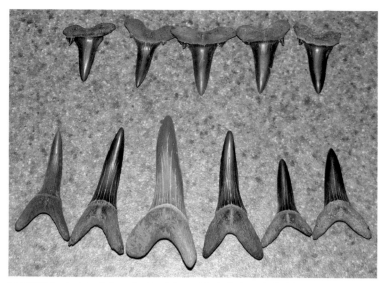

Fossil Goblin shark teeth from the Frankstown, Mississippi, geological site.

FICTION: Sharks do not get cancer, and taking shark cartilage is an effective treatment for cancer.

FACT: It has been known for quite some time that sharks do contract cancer. Many have been found with various kinds of tumors. There is absolutely no evidence that shark cartilage taken orally is effective for treating cancer in humans. Shark cartilage is nothing but modern-day snake oil.

FICTION: If there are dolphins around there will be no sharks.

FACT: Sharks and dolphins are not mortal enemies, although they have been portrayed as such on television and in the theater. If there are dolphins around there may still be sharks in the vicinity.

FICTION: Sharks have remained unchanged for hundreds of millions of years and are very primitive fish.

FACT: Although sharks have been on the earth for some 300 million years, they have changed significantly in that time. To consider sharks primitive would ignore the fact that they have very advanced reproductive and sensory systems as well as a number of other advanced characteristics.

FICTION: Sharks are not true fish.

FACT: Sharks are indeed fish. They are just unique fish.

SHARK BIOLOGY: STRANGER THAN FICTION?

Sharks can seem stranger than fiction and while I hope this book will dispel some of the myths concerning sharks, it is also worth mentioning some of the unique aspects of shark biology. The earliest sharks appear in the fossil record some 300 million years ago and by 150 million years ago, sharks very similar to modern species had become established. We know about sharks through the ages primarily from the large numbers of fossil shark teeth that can be found around the world and in many places around the Gulf of Mexico. I have personally found fossil shark teeth in south Florida (Venice Beach), in south-central Alabama (Little Stave Creek near Jackson), and in north Mississippi (Frankstown). Goblin shark (page 118) teeth are

Bull shark jaws from a specimen weighing approximately 300 pounds

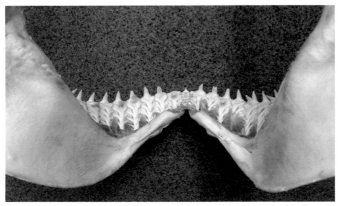

The view from the inside of a bull sharks jaws. Note the many rows of teeth on the inside of the lower jaw. This jaw contained a total of about 350 teeth.

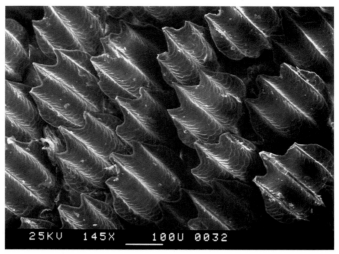

Shark dermal denticles magnified 145 times.

the most common fossil by far at the Frankstown, Mississippi, site. In fact, fossil shark teeth are frequently the dominant fossil in many ancient ocean deposits. The explanation for this is found in the oral biology of sharks. The shark jaw is literally a tooth "assembly line." As the shark grows and when teeth are broken during normal feeding, sharp new teeth are ready to roll into place. If the inside of the shark jaw is closely examined (I suggest you not try this on a live shark), hundreds of replacement teeth are easily observed. In one set of bull shark jaws, I counted about 7 teeth per row in fifty rows. A bull shark therefore sports 350 teeth at any one time. It has been estimated that a shark could go through thousands of teeth in a lifetime. It should therefore come as no surprise that the ocean's bottom is covered with discarded shark teeth.

Sharks and their relatives have scales unlike those of any other fishes. These unique scales are called **dermal denticles** or **placoid scales** and are barely detectable with the unaided eye. As a result a microscope is required to see them clearly. However, anyone who has ever had the chance to stroke a shark has felt the sandpaper-like texture of these scales. As a matter of fact, shark skin was once used as sandpaper in the furniture industry. The skin is so abrasive that when I have handled sharks for long periods of time without gloves, my hands felt raw and uncomfortable. If a shark brushes against you it can leave a nasty welt!

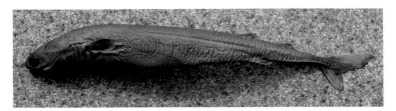

FEEDING

Sharks feed on a variety of prey using a variety of methods. Thankfully, the largest shark in the Gulf, the whale shark (*Rhincodon typus*, page 108), feeds on some of the smallest prey. These leviathans use their large mouths and specialized gills to filter plankton (tiny plants and animals) from the oceans. If these largest of fish were predators, swimming or even boating in the ocean would be a far riskier proposition. Another large shark that has not yet been recorded from the Gulf of Mexico but that will likely eventually be found there is the megamouth shark (*Megachasma pelagios*). This shark is likewise a filter feeder but apparently has **bioluminescent** (light-producing) tissue inside its cavernous mouth that may be used for attracting and concentrating its small prey. The vast majority of sharks prefer to feed on live or freshly killed prey. Most do not like decomposing flesh. As a matter of fact, there is some evidence that rotted shark flesh may actually repel some sharks and a chemical derived from decomposing shark was actually tested by the military as a shark repellant. Many of the bottom-dwelling species (e.g., nurse shark, dogfish) consume invertebrates such as squid, octopus, crab, shrimp, and lobster. One of the favorite prey of many sharks is other smaller sharks. However, sharks like most fish are opportunistic; that is, they will take a wide variety of prey if it presents itself.

Two images of the Cookie cutter shark, *Isistius brasiliensis*

Perhaps the strangest feeding habit of all the sharks is found in the cookie cutter shark (*Isistius brasiliensus*), a member of the dogfish family Squalidae. This shark is found in the deep ocean waters of the Gulf. It has very weakly developed fins and a relatively flabby

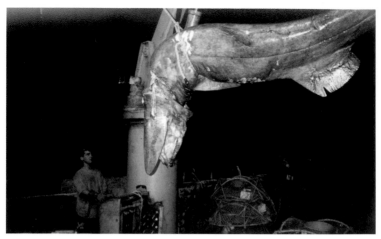

Goblin shark, *Mitsukurina owstoni*, captured 120 miles due south of Pascagoula, Mississippi

body, and it looks a bit like an Italian sausage with eyes. Although deceptively innocuous in appearance, it is actually a parasite that preys upon organisms much larger than itself. This species has thick fleshy lips that it uses to attach to the side of its unsuspecting prey. The shark then uses its sharp lower teeth to excise a "plug" of flesh, which it then consumes. Large scars, likely the work of this shark, have been found on the side of sperm whales, fish, and even on the rubber housing of underwater research equipment. These sharks are also bioluminescent, producing a green glow that may be used to attract unsuspecting prey.

Another shark that apparently has an unusual feeding method is also arguably the ugliest shark in existence. The goblin shark (*Mitsukurina owstoni*, page 118), a deep-water species, has jaws that are so loosely attached to the head that they look like they have fallen out altogether. The jaws are pincer-like and may be used to capture bottom-living organisms.

REPRODUCTION

Depending on the species, shark reproduction runs the gamut from egg laying to live birth. All sharks have internal fertilization. Unlike many other fishes, it is quite simple to tell a male from a female shark because all male sharks have **claspers** and the females do not. The claspers are simply modified fins and are located on the underside of the shark. Although all male sharks possess claspers, they

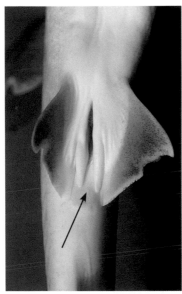

Claspers of an immature male shark   Claspers of a mature male shark

A bonnethead shark, *Sphyrna tiburo*, giving birth in the Florida Keys

grow rapidly when maturity is reached such that adult males can be identified by the length of the claspers.

For the vast majority of sharks, actual mating behavior and fertilization have never been observed. However, we have some information concerning the events that lead up to mating. In some species

Umbilical scar on a recently born blacktip shark pup

mating is relatively violent, with the males biting the females repeatedly between the first and second dorsal fins. Mating, the few times it has been seen, involves the male wrapping his body around the female while holding on to the pectoral fin with his mouth. While in this position the male inserts one of his claspers into the female's cloaca and inseminates her. The time from fertilization to birth, the **gestation period**, varies among species but is typically ten to twelve months.

Shark birth has been seen in only a few species. Thanks to Sea World of Florida, we were able to videotape bonnethead sharks (*Sphyrna tiburo*, page 84) giving birth. Gravid (pregnant) sharks

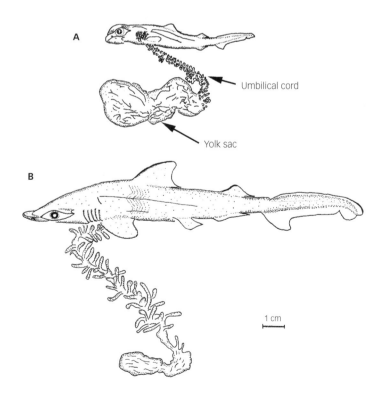

A

Umbilical cord

Yolk sac

B

1 cm

were held in captivity at a Sea World facility in the Florida Keys. When they were near term, we placed them in an observation tank and videotaped them as each **"pup"** (a newborn shark) was delivered. All sharks were born tail first and upside down. We described a "bat turn" behavior wherein the female swam rapidly and made a quick turn which facilitated the delivery of the pup. In fact, the pup literally tumbled through the water because of the speed at which it was thrown from the mother.

In those sharks that have live birth, an **umbilical cord** is present in some species. The umbilical cord attaches the embryo to the mother and is broken during birth. The presence of an **"umbilical scar"** can be used to determine the recency of birth in some species. A pup will keep a very prominent scar for several days after birth. The scar can be seen on the underside of the shark just ahead of the pectoral fins.

Perhaps the oddest reproductive method of all is found in the sand tiger shark (*Eugomphodus taurus*, page 105). The mother shark produces a number of eggs that are fertilized and implanted in the uterus. As the embryos develop, one embryo in each uterus (sharks have two uteri) gets the upper hand on the other embryos and eats them (referred to as **intrauterine cannibalism**). Sibling rivalry at its earliest and most lethal! The mother shark then produces more eggs that move into the uterus and the developing embryo eats these as well. For obvious reasons, a maximum of only two pups are born at a time. One of the first scientists to experience this kind of cannibalism was actually bitten by an embryo of a sand tiger shark. As the

A sand tiger shark, *Eugomphodus taurus*

Cuban dogfish, *Squalus cubensis*, with embryos

story goes, the unborn shark bit the scientist on the hand while he was conducting a uterine examination of the mother. That must have come as a real shock!

Most sharks produce a few well-developed young but some have been found containing as many as one hundred embryos. In the most common sharks of the Gulf of Mexico, six to ten pups is typical. When shark pups are born they are immediately ready to fend for themselves. The mother shark plays no part in their upbringing. As a matter of fact, sharks are cannibalistic and may actually eat their young if given the chance. This does not appear to happen during birth as the mother shark may stop eating in preparation for birth, preventing her from making a snack out of her pups.

The Cuban dogfish (*Squalus cubensis*, page 103) is found in the southern Gulf and is one species in which females do not have an umbilical connection to their pups. Instead, the developing eggs are very large and are held within the body of the shark until they hatch. When the embryos are examined prior to birth, a large amount of egg yolk contained within a sac is found attached to the embryo.

SHARK SENSORY SYSTEMS

Sharks have some of the most advanced sensory capabilities of any animal, yet we know very little about the sensory ability of the vast majority of species. Much of the information that has accumulated has come from studies of shark attack conducted by the U.S. Department of Defense, Office of Naval Research. The navy was very interested in shark sensory capabilities for the purpose of developing an effective shark repellent. Although their efforts to develop a shark repellent resulted in mixed success, our understanding of shark vision, olfaction (smell), and hearing has been greatly advanced.

Vision in the ocean presents a different set of problems than vision on land. Water visibility can be zero, but even at its greatest it probably does not exceed much more than one hundred feet. Sharks have rather acute vision that is effective at distances of ten feet or so. Some sharks have been shown to have both **rods** and **cones** in their **retinas**. What this means is that they can evidently see some color (cones allow for color vision). The eyes of sharks that have been examined are most sensitive to blue and green light. This makes sense because red and yellow light are quickly attenuated by water such that blues and greens are the only light available at depth. Additionally, the light produced by organisms (bioluminescence) that inhabit the oceans is likewise blue and green. There are a number of sharks in the dogfish family (page 102) that possess the ability to produce light. All of these bioluminescent species are found in the midwater depths of the deep sea. The light-emitting organs are called **photophores**, and they are typically located along the bottom and sides of the shark. Bioluminescent sharks found in the deep water of the Gulf of Mexico are small, generally black in color, and, as is typical of dogfish, have sharp spines in front of the dorsal fins. The green dogfish (*Etmopterus virens*), one of these glowing dogfish, is apparently common in the very deep waters of the Gulf of Mexico.

An interesting feature of the shark eye is the presence of a protective covering called the **nictitating membrane**. This membrane normally rests below the eye and is raised when the shark feeds. The membrane protects the eye from being damaged by thrashing prey. The shark also has a unique reflective coating behind the retina (the light-sensitive layer) of the eye. Many nocturnal (night-active) animals have a similar layer called a **tapetum lucidum**. Cat fanciers may have observed the reflective layers of the eyes of their pets

A bioluminescent dogfish from the deep sea

Bull shark's Ampullae of Lorenzini

glowing at night when a light is shone on them. The tapetum reflects light back through the retina to increase the light sensitivity of the eye. Deep-ocean sharks have a very strong tapetum such that the eyes glow an eerie green at night.

If you examine the surface of a shark very closely, you will observe tiny pores along the body and especially concentrated in the head region. These pores are called **Ampullae of Lorenzini** and are the openings of a canal that leads to specialized sensory organs that allow sharks to detect electric fields. If you have ever "felt" the static charge of clothing that has just come from the dryer, then you have an idea of what a shark must sense with its Ampullae of Lorenzini. However, the shark's sensitivity to electric fields is amazing. Sharks can actually detect the extremely small voltages produced by the contraction of the heart of a fish or crab buried in the sand. So how are these sensory organs used by sharks? When a shark is searching for prey it may swim along the bottom, using its Ampullae to sense any electric charge coming from the sand. If it detects an electric field it will attack the bottom and, hopefully, find a fish. The hammerhead sharks (page 79) are particularly adept at feeding on stingrays, a fish that likes to bury in the bottom. It has been suggested that the extreme expansion of the head in these sharks allows for greater Ampullae coverage area of the bottom (sort of like a mine detector!). As the shark swims along the bottom it may better detect these hidden food items. The electrosensitive ampullae are also effective in prey location at night when vision is less useful. Although the electrosensitive ability of sharks is acute, this sensory system is only effective at a few feet or so at best. Long-distance detection of prey is best performed by olfaction and hearing.

A very large percentage of the shark brain is devoted to the processing of olfactory information. This is testimony to the importance of the sense of smell in prey location. Sharks have a very acute sense of smell, and chemicals dissolved in water can be carried for great distances. It is entirely conceivable that a shark could detect a potential source of food from many miles away. Certainly fishermen take advantage of the shark's acute sense of smell when they "chum" (see Shark Fishing and Fisheries, page 32) for sharks by ladling ground fish into the water.

Finally, the sensory system that no doubt is most effectively used at long distance is hearing. Sharks can detect vibrations in the water by both the ear and the lateral line system. The ears of sharks are most sensitive to low-frequency vibrations. These are precisely the kinds of vibrations that are produced by a struggling fish. It has been found, through experimentation, that sharks can be attracted from a great distance if low-frequency sounds are played through a hydrophone (an underwater loud speaker). It is interesting that pulsed sound is more effective at attracting sharks than continuous sound. The **lateral line** is also used to detect vibrations (sound) in the water. This system is hardly noticeable but occurs as a series of pores arranged in a line that runs along the flanks of the shark.

SHARK ATTACK

I probably get more questions from the general public about this topic than any other. What are my chances of getting bitten by a shark? Is it safe to let my kids swim at the beach? I even had a person ask me once if it was O.K. to swim in a local freshwater lake! The fear of shark attack is very much blown out of proportion. If the average person feared driving their automobile (a far riskier proposition) as much as they apparently worry about shark attack, then we could eliminate our dependence upon foreign oil overnight. To emphasize this point I have assembled a list of circumstances that, in all probability, would be more likely to occur than being bitten by a shark.

BEFORE BEING BITTEN BY A SHARK YOU WOULD BE MORE LIKELY TO

1. die in an automobile crash (40,000 deaths each year),
2. die from flu complications (36,000 deaths per year),
3. die in a firearms-related accident (30,000 deaths per year),
4. die from prescription drug errors (7,000 deaths each year),
5. die in workplace injuries (1,500 deaths each year),
6. be struck by lightning (100 deaths each year),
7. die from a bee or wasp sting (25 deaths each year), or
8. die from a snake bite (15 deaths per year).

Fewer than one shark attack per year proves deadly to humans.

The point is that you have to put things in perspective. Sure, you can totally eliminate your chance of being bitten by a shark by never swimming in the ocean, but is it realistic to deprive yourself of one of life's great pleasures on the very, very low chance that you might encounter a shark?

SIMPLE RULES TO HELP PREVENT SHARK ATTACK

While the likelihood of being bitten by a shark is very low, there are some simple rules to follow that can further reduce your chance of encountering a shark:

1. Do not swim early in the morning or late evening and certainly not at night. We know that many kinds of sharks are most active during dusk and dawn and are feeding during those hours. I do not let my children swim before 9:00 A.M. and not after about 6:00 P.M.
2. Do not swim in an area where fishermen have bait in the water, particularly near fishing piers or jetties or, more important, if fishermen are using highly scented baits that could attract sharks from a distance.
3. Do not swim alone. At the very least this spreads out the already very low probability that you will encounter a shark. Also, a swim partner is a necessity anytime you are swimming just in case something unforeseen happens.
4. Do not swim in murky, highly stained, low-visibility waters. The chance of a shark mistaking a hand or foot as a possible food item is higher in waters that are cloudy/murky.
5. Know the area you are swimming in. Ask local residents if there are areas known for high shark numbers and stay out of those areas.
6. Obviously, do not swim if you have an open wound or if there is a chance that blood may enter the water.

7. There is some evidence that sharks congregate around passes, areas where water flows between islands or out of a bay or sound. Avoid these areas.
8. Do not swim in schools of baitfish. Some sharks feed by rushing into a school of baitfish and biting at whatever is there. If you get between the shark and its food, you may be accidentally bitten.
9. Do not swim far from the beach. The deeper the water, the greater the chances of encountering a large shark. Additionally, if you happen to get bitten far from the beach, it will be harder for help to reach you.

SHARK CONSERVATION

Why should anyone care about conserving sharks? The easiest answers are practical in nature. Sharks are a valuable, renewable resource. Shark meat has gained in popularity and is enjoyed by many people worldwide. You can be sure that those folks would not want to see sharks disappear from the oceans. Also at issue is the idea that there is a "balance of nature" in the oceans. Although this is an oversimplification, it is true that sharks play an important role in regulating the populations of many different kinds of marine organisms. For example, sharks may help keep less "desirable" species (such as hardhead catfish) in check and prevent them from competing with other popular sport/commercial species (such as redfish, snook, and sea trout). With sharks removed, the less desirable species could increase in number and compete with the more popular sport or commercial species, thus changing the marine community in unpredictable ways.

Another practical consideration for conserving sharks is the potential for new product development. Having many species on the earth presents a potential cornucopia of new drugs, pharmaceuticals, cosmetics, and so on. The extinction of even a single species means the loss of a vast amount of information that could potentially alleviate much human suffering. For example, some kinds of shark liver oils have been investigated for their use in preventing disease. **Squalene**, derived from the liver oil of deep-sea sharks, may indeed have human health benefits as a dietary supplement. Allowing a species to disappear from the earth is like discovering some ancient texts in a language that we cannot yet decipher and then burning them all. Who knows what mysteries might have been solved or what inspiration might have been provided by the words written there!

A final consideration for conserving sharks is an aesthetic one. There are some people, myself included, who take comfort in knowing that there are still places on the earth where wild animals roam. I feel that the world would be a less interesting place if sharks were not in it. I believe the earth is diminished each time we lose a species. For this reason alone we must strive to preserve all of our living marine resources.

SHARK FISHING AND FISHERIES

As noted above, I take stewardship of the earth's resources very seriously, but I am also a fisherman and have spent countless hours fishing for sharks in the Gulf of Mexico. I have no moral problem with legal shark harvest, although I practice only shark catch and release. I do have a serious problem with people who do not obey fishing regulations. It is these "outlaws" who give fishers a bad name. Prior to engaging in any fishing activity, it is very important to familiarize yourself with the fishing regulations for your area. This information can be easily obtained from your state Wildlife and Fisheries Department. If you have any question about fishing regulations in the Gulf area, refer to table below for contact information.

GULF STATES WILDLIFE AND FISHERIES DEPARTMENTS

Department	Phone	e-mail
Alabama Marine Resources Division	251-861-2882	dcnr.state.al.us/
Florida Fish and Wildlife Commission	850-487-0554	floridaconservation.org/marine/
Louisiana Dept. of Wildlife and Fisheries	225-765-2383	wlf.state.la.us/apps/netgear/
Mississippi Department of Marine Resources	228-374-5000	dmr.state.ms.us
Texas Parks and Wildlife Dept.	800-792-1112	http://www.tpwd.state.tx.us/

There are several species of sharks that are protected in federal waters and in some state waters. Sharks that should not be taken in

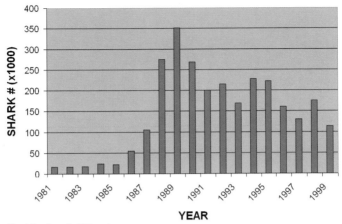

Shark landings in U.S. waters

the Gulf of Mexico are the white shark, dusky shark, sandbar shark, and great hammerhead shark. The whale shark is likewise protected, but it is unlikely that this species would be encountered in shallow water. My recommendation is that if you are in doubt about the identity of a shark or the regulations regarding it then release it.

SHARK FISHERIES

Sharks have been harvested at one time or another for their flesh, hide, fins, liver oil, jaws, teeth, and various other parts. Over the past fifteen years, there has been a tremendous increase in the numbers of sharks harvested by both recreational and commercial fishers in the Gulf of Mexico and the Atlantic Ocean. This increase in harvest has occurred because of the growing acceptance of shark meat for the table and the increased demand in the shark fin market. Shark fins are used to make shark fin soup, which is very popular in some Asian countries. This can be such a lucrative fishery that less than scrupulous fishermen have captured sharks, cut their fins off, and thrown the rest of the shark overboard. Remarkably, some of these "finned" sharks were later recaptured. The fact that some sharks were able to survive with no fins is a testament to their resilience. The practice of shark finning has been banned in state and federal waters. Unfortunately, after a peak in U.S. landings in 1989 the trend in shark landings has been down. This suggests that shark numbers are on the decline, and some species may be dangerously depleted and close to extinction.

With proper management, some kinds of sharks may be commercially harvested; however, as a group these animals are very poor subjects for exploitation. Many sharks have very slow growth, have low rates of reproduction, take years to reach maturity, and live to be very old. These characteristics can be a recipe for disaster if the number of animals captured is not closely regulated. For this reason it is important that scientists understand as much as possible about sharks (see Shark Biology section) so that wise management decisions can be made.

HOW TO SAFELY CATCH, HANDLE, AND RELEASE A SHARK

For the sport fisher, capturing a shark on hook and line can be a thrilling experience. In the Gulf of Mexico, species that are very popular as sport fish include the blacktip, the Atlantic sharpnose, the finetooth, the spinner, and the mako, although the mako and spinner are not particularly common. Smaller species of sharks (e.g., finetooth, sharpnose, and juvenile blacktip) can also be great fun on hook and line and are actually better to eat than larger species. My experience is that large sharks have a very "grainy" textured flesh that is not particularly appealing, another reason for releasing larger sharks.

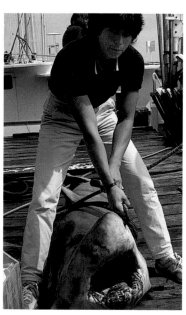

Fishing for small sharks requires little specialized equipment. Before going on any excursion on the ocean be sure you have made someone on land aware of when you will depart, where you will be fishing, and when you will return. Because I believe that large sharks should not be harvested, I will discuss only sharks less than about five or six feet long. A five-foot shark is an impressive fish and is plenty of animal for the average fisher to handle. If you happen to hook a very large shark my recommendation is to not try to land it but to cut the line and release it. Handling large sharks is dangerous and should be attempted only by

experienced fishermen. It is just not worth the risk of having your favorite rod broken, having your boat trashed, or getting bitten by a big shark thrashing around on deck.

To catch even a small shark requires a length of steel leader that is at least as long as the shark that you are attempting to catch. I prefer solid steel leader, not the woven steel kind, because it seems that I always manage to get some of the wire stuck in my hand when the braids unravel. Never use prefabricated leaders. Make your own so you will only have yourself to blame if your leader breaks. I prefer at least 40- or 50-pound test line on a Penn 4/0 reel with a stiff boat rod. With this combination you can easily land sharks up to five or six feet long. In fact, this tackle is actually overkill for these small sharks; however, my strategy is to get the shark to the boat as quickly as possible, decide if it is to be kept or not, and if not, release it while it is still very lively. This gives the shark the best chance of surviving after capture. If you use lighter tackle, there will be more time required to land the shark, the shark will be more stressed when landed, and, as a result, it will have less chance of surviving if released. You do not need a huge hook for small shark fishing. For small sharks I use a standard, straight shank, 4/0 or 5/0 hook. The Japanese circle hook is also popular for shark and may provide better catches. However, I am concerned that the circle design may actually cause more tissue damage and may contribute to greater mortality in released fish.

A tiger shark swims at the water's surface.

A LESSON ON USING ENOUGH STEEL LEADER

Back in my early days of shark fishing, when no one was particularly concerned about shark conservation, my buddies and I were shark fishing out of Clearwater, Florida. When we were rigging up for shark I made up a leader about five feet in length, thinking this would be plenty. We spent the entire night fishing and chumming and caught and released many small sharks. However, just as the dawn was breaking and with my buddies sacked out down below, I thought I could see a shark fin cutting through the water just outside the lights cast by the boat. As the sun came up I got a good look at the fin, and it was indeed a shark. I could not tell the size of the fish but the bits of algae and debris that were caught on the fin told me it was a tiger shark. For some reason the fins of tiger sharks will sometimes have bits of debris or algae trailing from the rear edge. I have seen this on a number of tiger sharks. Tigers like to swim Hollywood style, just under the water with their dorsal fins slicing the surface. I woke my friends and we watched as the shark circled closer and closer to the boat. Our first good look at the shark was when it swam underneath the boat from "stem to the stern." The three of us were standing looking over the transom of the boat as the shark seemed to materialize from underneath. It was one of those "we gotta get a bigger boat" kinds of experiences. The shark was at least five hundred pounds and looked to be ten or twelve feet long, and that may have been a conservative estimate! After a few more minutes of circling, it bit one of the baited hooks and then spat it out. It immediately went to the bait on the end of my line and grabbed it. I set the hook and then moved to the bow as the shark literally dragged around a twenty-four-foot boat for about an hour. Unfortunately, every time the shark would make a run I could feel the tail of the shark slapping my fishing line. Eventually, this constant abrasion caused the monofilament to snap. The shark was much longer than the steel leader I had prepared, and the shark "tail whipped" the monofilament line. I have to admit that later on I was secretly glad that we did not catch that shark. I like to think that she may still be prowling the waters around Tampa Bay.

The three most important factors to consider when shark fishing are location, time of day, and season. In my experiences in the Gulf of Mexico, the best places for shark fishing are around passes and in bays and sounds. Structures such as natural or artificial reefs, gas or oil rigs, or piers and bridges attract small fish, which in turn attract sharks. The gas and oil rigs offshore of Alabama, Mississippi, and Louisiana are good places to catch sharks. Likewise, the Mississippi Sound around Dauphin, Ship, Horn, Round, and Cat islands are also shark havens. The northern Gulf of Mexico is known for large concentrations of small blacktip, sharpnose, and finetooth sharks, and these may be encountered in almost any area. In Louisiana, the Chandeleur Islands are a great place to catch blacktip sharks and, in fact, they can be real nuisances for those trying to catch sea trout or redfish. The shallow waters around the Mississippi River delta tend to be very low salinity and are good habitat for juvenile bull sharks. One of the greatest concentrations of sharks (finetooth, blacktip, bonnethead) I have ever witnessed was in the Crooked Island Sound near Panama City, Florida. Another exceptional spot for sharks are the waters around Cape San Blas, Florida, particularly the Indian Pass area. Along the central coast of Florida the old Sunshine Skyway bridge in Tampa Bay is a great place for shark fishing. I have caught and released many blacknose sharks while fishing at night from that bridge. Another good spot in Tampa Bay is the shallow flats just off of Coquina Key. This is an excellent spot for small blacktip and bonnethead sharks. Further south I have caught many sharks, mostly bonnethead and blacktip sharks, in the Florida Keys around Long and Marathon keys. Bonnethead sharks in particular can be found in large concentrations on the grass flats in the Florida Keys and in many places in Florida.

Although it is possible to catch sharks at almost any time of day, it is well known that many species actively hunt for food at night, in particular around dawn and dusk. I have always preferred night fishing for sharks because it is considerably cooler and more comfortable, and the sharks bite better at night. Another consideration is that sharks seem to suffer less capture stress when caught at night, which means their chance of survival upon release is greater. If you choose to fish at night be sure to review all of the safety considerations for venturing out on the ocean after sundown. The absolute worst time to shark fish is typically midday, although this can change depending upon season and tidal stage. Our studies in the Gulf have shown that during the summer, sharks do not prefer to be in very warm shallow water during the middle of the day, so fishing during this time may not be successful. I have always preferred to fish on an incoming tide, and many other shark fishers will agree with this. If there is considerable freshwater input into your area because of an outgoing tide, then this may drive sharks further offshore.

The time of year that is best for shark fishing will change depending upon your location in the Gulf. In the southern Gulf, sharks may be encountered throughout the year. There is little seasonal cooling so sharks stay inshore nearly year-round. For example, in south Florida and the Florida Keys, I have captured sharks year-round in shallow waters, although catches are lower during the winter months. Further north, the arrival of winter forces sharks out of inshore waters, and from about December to March shark fishing is very poor to nonexistent. In the northern Gulf shark fishing is best from May to October. Along the central Florida Gulf coast the situation is similar with shark catches being low in the winter months but higher in spring, summer, and fall.

Once you have chosen the appropriate tackle and you have anchored in your fishing spot, it is time to consider how you are going to fish. I suggest using fresh whole fish as bait and hooking the fish through the head or tail. If you use soft baits like menhaden or mackerel, make sure you put the hook through a fairly substantial part of the fish so the bait does not fall off. Any kind of fresh fish will do. Bonito is a popular shark bait because the flesh is very bloody and it is a very firm bait. It is best that you do not allow your bait to lie on the bottom because it will be quickly consumed by crabs and small fish. I find that floating the bait a few feet off the bottom by attaching a balloon, a large fishing float, or a piece of styrofoam to the line will reduce some of the problem of having your bait stolen. However, be sure to retrieve the balloon or styrofoam float so that you are not polluting the environment. In a strong current the bait will tend to rise to the surface, so it may be necessary to put a few ounces of weight on the line to hold the bait near the bottom. Allow the current to float your bait away from the boat. I like to float the bait about fifty feet or so down current. Place your reel in free spool and engage the ratchet. The sound of the ratchet will alert you when the shark takes the bait. Either hang onto the rod (which is my preference) or place it in a rod holder.

BE CAREFUL OF THE END WITH THE TEETH

While catching sharks near Ship Island, Mississippi, for a television program, a finetooth shark (page 64) that I grabbed by the tail, grabbed me by the leg. Finetooth sharks are very tenacious and when they latch onto something they do not like to let go. On video (our research project was being filmed for later broadcast), it appeared that the shark had me by the flesh. Fortunately, I was wearing long pants over rubber knee boots, and the shark managed to get a purchase only on my clothing and boot. I was able to watch the whole thing on video and it was somewhat comical. My graduate student tried to help but she was, understandably, afraid of losing a finger. Fortunately, the shark finally let go voluntarily. Later, upon examining the boots, I saw that the shark's teeth sliced completely through the thick rubber. If I had not been wearing those boots it would have done real damage to my leg. The lesson here is that if you have a shark by the tail, be aware of the end with the teeth.

EQUIPMENT RECOMMENDED FOR SHARK FISHING

Tackle box
Fishing rod/reel
Device for making chum
  or frozen chum
Mesh bag
Gaff
Fire extinguisher
Crimps
Cotton gloves
Wire cutters
Steel leader

Hooks
Weights
Bait
First aid kit
Cell phone and emergency
numbers
Navigational charts of
  the area
Emergency flares
Personal floatation devices
Weather radio

While your bait is in the water, you may want to chum. Chum is basically ground fish that can be purchased frozen or that you can make yourself using a meat grinder or one of the commercially available chum grinders. If using frozen chum, place the block in a mesh bag and lower it over the side of the boat until it is just covered with water. Currents and the motion of the boat will help disperse the scent. Menhaden or mackerel make good chum because their flesh is very oily and the scent will carry a great distance. The chum "slick"

DON'T EAT THE BAIT!

I've spent countless hours "soaking bait" in the waters of the Gulf of Mexico and have captured literally thousands of sharks over the past twenty-five years. Some of my earliest shark fishing experiences took place in the late 1970s in the vicinity of Dauphin Island, Alabama while a research fellow at the Dauphin Island Sea Lab. A typical day of shark fishing would begin by preparing the boat, gathering all the necessary accouterments, and purchasing bait and ice at the local seafood market. I frequently used mullet for bait because it was cheap and, if Neptune did not smile upon you that day, you could always have it for supper. On one particular summer day, after several scorching hours of fishing, I returned home with no sharks and most of my bait untouched. Late that evening, after cleaning up, I sat down to a sumptuous plate of fried mullet. I turned in at about 10:00 and as soon as my head hit the pillow, despite the fact that I was dog tired, my mind raced over the events of the day. The school of whale sharks that had almost capsized the boat, the Pelican that spoke perfect German who selfishly drank my last Sprite, and the band of angels that played Hank Williams tunes on their harps while flying around a nearby shrimp boat. I jumped from the bed and turned the lights on to find that my sheets and pajamas were soaked with perspiration. I shrugged it off as a very vivid nightmare, moved to the living room where it was cooler, and lay down on the couch. As soon as I closed my eyes my mind raced out of control. I was caught up in a cyclone of strange emotions and then the hallucinations began again. Scarlet monkeys, sharks swimming through fountains of orange light, multicolored lightning bolts flashing across the room. This continued for the entire night. Most of the hallucinations I have no memory of. At the worst part of the incident I was close to calling someone to drive me and my pet unicorn to the emergency room. The visions finally subsided in the wee hours of the morning, leaving me completely exhausted and terribly frightened. Concerned that I would be branded a lunatic and quite frankly afraid that they might be correct, I told no one of this incident. The next evening I was afraid to go to bed for fear that the visions would return. Fortunately, they did not. Over time, my memory of the incident faded and it would be almost a year before I discovered the cause of my presumed brush with insanity. While perusing an obscure publication called the *Journal of Tropical Medicine*, I happened upon an article entitled "Hallucinatory Mullet Poisoning." It was one of those "Eureka!" moments. The symptoms described were exactly those that I had experienced. Most cases of the poisoning are reported from the tropics where it is extremely rare. It is unheard of in the United States. The symptoms, which develop in five to ninety minutes, include sweating, weakness, hallucinations, and/or vivid nightmares. One species of Hawaiian fish is even called the "nightmare fish" because it is more likely to contain the hallucinogenic toxin.

You can imagine my relief to discover that I had not lost my mind and I was able to tell my friends about the experience. One of my more liberal-minded friends, I think we all know the type, asked me why I didn't just enjoy the experience. I responded that maybe I could have looked upon it more objectively if I had known the cause of this "altered consciousness" and I asked him how he would feel if he suddenly began to hallucinate for no apparent reason. I also explained that watching snakes crawling across the ceiling of my bedroom was not my idea of fun! Two things stayed with me from my experience with hallucinatory mullet. One was an anxiety about eating mullet that lasted for several years. The other was some small appreciation for what it must be like to lose one's mind. I dearly cherish my sanity, almost as much as Stubby, my pet unicorn.

is the oil that has been released from the fish flesh. You can actually see the slick on the surface of the ocean as it moves down current. Remember if you change locations, you will break the chum slick and have to start over again. However, after a number of hours with no bites you may have to reel 'em up and look for greener pastures.

Typically, when a shark takes a bait it grabs it and runs. You of course want to make sure that you have your drag adjusted properly before this happens. As the shark swims it will get a better purchase on the bait, so allow it to run for a few seconds; however, do not let it run for too long because it may swallow the hook. After a few seconds, engage the spool, carefully reel up any loose line, and pull back suddenly on the rod to set the hook. If the drag is set too loosely, there will not be enough resistance to set the hook properly; if it is set too tight, you will perhaps break the line or your rod. At this point the shark will likely make another hard, fast run so be prepared. The line should pay out from the reel with enough drag to bend the rod and to keep the line very tight. This resistance will slow the shark and allow you to reel up line. Keep the rod tip up so the rod itself absorbs the force of the pulling shark. Use patience and gradually work the shark up to the boat. The shark will likely make a second run when it sees the boat, but it will not be as vigorous as the first. Work the shark to the boat once again and prepare to either release the shark or land it.

Certainly a consideration of utmost importance in shark fishing is how to handle the animal once you have captured it. I have handled literally thousands of sharks and after a day of capturing, tagging, and releasing sharks, my bare hands can feel like chopped liver (remember those rough, sandpaper-like scales). So my first recommendation is to get yourself a pair of soft cotton gloves, like those used for gardening, and put them in your tackle box. Before handling a shark be sure to wear these gloves. I suggest that if the shark is greater than about five feet in length, cut the line and release it; do not bring it into the boat. A live, large shark thrashing around on the boat, biting at anything within reach is too dangerous for the average fisherman. NEVER try to remove the hook from a live shark with your bare hands, no matter the size of the shark. A small blacktip shark, only about three feet long, took a good chunk out of my thumb when I reached to grab the shank of the hook that protruded from the shark's mouth. It is best to use long-handled pliers for hook removal on small sharks and on big ones simply cut the line as close to the shark as is safe but do not place your hand near the shark's mouth. Do not try to remove the hook from any shark that is gut hooked or hooked in the gills. You will simply do more damage to the animal. In these cases it is best to cut the line and release the animal. The hook will corrode away in a relatively short period of time.

If you have no intention of keeping any sharks, then take along your camera, snap a photo of the shark while still in the water (particularly if it's a large one), and cut the line well above the mouth to prevent getting bitten. Releasing a shark that has been brought onboard the boat is not particularly difficult. However, caution should be exercised to prevent injury to yourself and to the shark. Use of a very large dip net is highly recommended for both landing the shark (if small enough) and for returning it to the water. Often the shark will bite the webbing, so a good-quality landing net is recommended, although over time the teeth and rough scales will take its toll. If the shark takes a big bite of netting do not try to pull the netting from its mouth. This will only ruin your net. Simply hang the net and shark over the side into the water, and wait for the shark to voluntarily release from the net. This typically only takes a few seconds or minutes. If you have to handle the shark to get it overboard, then make sure you have a good purchase on it. For very small sharks you can grab it by the tail and quickly place it overboard, being careful of the end with the teeth.

The Skates and Rays

SKATE AND RAY BIOLOGY

Although closely related to the sharks, there are fewer species of skates and rays in the shallow Gulf waters but if one ventures into the deep waters, a great many species of skates can be found. In the Gulf of Mexico there are about twenty-five common species of skates and rays, all contained within the group referred to as the **Batoidea** or **Batoids**. The Batoids are an interesting group that includes small butterfly rays that look like a delta-wing airplane, sharklike sawfish, and colossal manta rays exceeding twenty-five feet in wing span and large enough for a scuba diver to ride upon. The Batoids are divided into two large groups: the skates and the rays. While most people are familiar with stingrays, fewer have had the opportunity to see a skate. Skates look very much like rays but they have no stinging spine and the tail is much more developed; stingrays often have a long whiplike tail. Skates may possess large thorns on their body and/or tail but these thorns are relatively harmless. At first glance the sawfish (page 123) and guitarfish (page 126) appear very different from either skates or rays, but closer examination reveals many of the Batoid characteristics.

Skates and rays are closely related to sharks in a number of ways. Like sharks, skates and rays have a skeleton that is made entirely of cartilage. All sharks, skates, and rays have internal fertilization and produce large, well-developed offspring that are ready to fend for themselves. Like sharks, many skates and rays have live birth but others lay eggs.

FEEDING

Most of the skates and rays feed upon a variety of bottom creatures: shrimps, crabs, oysters, clams, and the like. Their teeth, unlike most sharks, are adapted for crushing and grinding, not for cutting. The

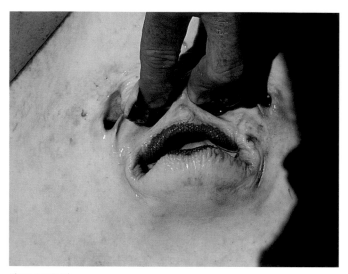

Skate/ray teeth

teeth are flattened and platelike and arranged in rows. Some species of rays, however, are filter feeders. The manta and devil rays (page 159) use the hornlike extensions on their heads to direct small organisms into their mouths. These organisms are strained from the water and then swallowed. Filter feeding, in this manner, requires constant forward movement and this group of rays is adapted for continuous swimming.

REPRODUCTION

The rays have live birth whereas the skates lay eggs. Skate egg cases are variously shaped depending upon species and are sometimes called **mermaids purses**. The most common egg cases found along Gulf of Mexico beaches are those of the clearnose skate. The clearnose skate egg case is dark brown to black and may have long stiff tendrils at each corner. The tendrils anchor the egg case to the ocean bottom, securing the egg during embryonic development and when hatching. The empty egg cases can sometimes be found cast upon the beach particularly after storms. Very rarely, an egg case with a live embryo inside may be found. I have placed the egg cases of this species in my home aquarium and watched the embryos hatch. As a matter of fact, the embryo can be carefully removed from its case, placed in a drinking glass in an aquarium, and the miracle of

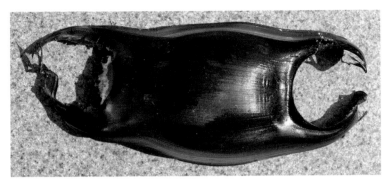

A skate egg case, sometimes called a *mermaid's purse*

skate development observed first hand. Mating behavior has been observed in several species of skates and rays, and it is similar to the behavior seen in sharks. There is variation in behavior depending upon species, but prior to and during mating, males may bite the females on the edges of the pectoral and/or pelvic fins. In some species there is simultaneous biting of each other's pectoral fins prior to and during mating.

DEFENSE

A stingray spine

Compared to their close relatives the sharks, the skates and rays are almost defenseless fishes. However, some interesting means of defense have developed in this group. The skates must rely on simple camouflage, using their coloration to blend into the background or burying in the bottom to avoid being detected by predators. The rays have one or more venomous spines on or near the base of the tail. The stingrays have a muscular, whiplike tail that can be lashed about as an effective weapon. More than a few beach-goers have had the misfortune of stepping on a buried stingray and receiving a very painful, serious wound (see section on stingray wound treatment). The stingray spine is dosed with venom, and although not normally lethal, I am told it is extremely painful. It is not uncommon to find sharks, particularly hammerhead sharks, with stingray spines in the stomach and even imbedded in the jaws.

Of all the rays in the oceans of the world, the electric ray's ability to defend itself has to be the most interesting. Its electric ability is famous (or infamous). The electric ability of these fishes has been known at least since the first century A.D. Ancient Roman physicians prescribed the electric discharge of the torpedo ray for a variety of illnesses, including gout and headache. They were even used for a sort of primitive electroshock therapy for treating mental illnesses. The scientific name of the two species in the Gulf of Mexico indicates their unusual abilities. The name *Torpedo* (page 130) comes from the Latin *torpere*, which refers to a state of mental or motor sluggishness or inactivity. Likewise, the name *Narcine* (page 129) comes from the Greek word *narke*, which means to numb.

BEHAVIOR

Anyone who has ever seen a manta ray or stingray when it is "on the wing" cannot help but be impressed with their graceful movements. The beating of their oversized pectoral fins gives the impression that they are flying through the water. Although graceful in swimming, they can nevertheless generate a great deal of thrust when startled and will disappear in the blink of an eye. This flapping style of swimming that is found in many of the species of skates and rays has earned them the scientific name Batoidea or Batoids. Depending upon the species, these fish may be observed lying on or buried in the bottom, swimming anywhere from just above the bottom to just below the water's surface, and even making spectacular leaps from the water.

STINGRAY WOUNDS: HOW TO PREVENT AND TREAT THEM

Stingrays are common inhabitants along Gulf of Mexico beaches, and during certain times of the year there can be large concentrations of rays in shallow waters. In many coastal communities, flags are posted along beaches to alert bathers of the presence of rip currents, jellyfish, stingrays, and other potential hazards. When stingrays are present, it is important to take precautions to prevent stepping on them. When wading in shallow water it is advisable to do the **stingray shuffle**. The stingray shuffle is simply dragging your feet along the bottom when wading rather than taking steps. If a stingray is lying on the bottom or is buried in the sediment and you drag your feet while walking, then your foot will strike the side of the stingray and the ray will immediately dart away with no harm to you or the ray. However, if picking up your foot and planting it ahead of you, there is no guarantee that you will not plant your foot right on top of a stingray with disastrous results.

If you are unlucky enough to be "stung" by a ray then it is **very important to get medical attention as soon as possible.** While for most people a stingray wound is not life threatening, an individual sensitive to the venom that is found on the spine could have serious problems. If immediate medical attention is not an option, the wound can be treated by immersion in water that is as hot as tolerable (110 to 115°F) but not so hot as to cause tissue damage. The stingray venom is a protein that is quickly inactivated with high heat. Those who have tried this treatment have reported that it works amazingly well. The immersion should last for thirty to ninety min-

THE SHOCKING ELECTRIC RAYS

Electric rays can be a real surprise to the unsuspecting. In one author's report, a fisherman trained his dog to locate and retrieve flounder from shallow water. The dog inadvertently grabbed an electric ray, and of course the fish defended itself as only electric rays can. The poor dog ran yelping into the forest and refused to fish from that day forward! My infatuation with this species has clouded my judgment on at least a couple of occasions. While sampling sharks in the northern Gulf of Mexico using a net, we pulled a very lively, lesser electric ray into the boat. I immediately pounced upon the animal to untangle it from the net, photograph it, and return it to the water as quickly as possible. My thought was that I could gently cradle it in my hands and put it over the side of the boat. As soon as I got both hands underneath the ray, it sent all of its 200(?) volts up my arms. To say it surprised me is an understatement. My right arm was numb and then very "tingly" for several minutes afterward. Thinking that the ray must be spent, I tried again to gently return the animal to the water with exactly the same results. I then instructed one of my graduate students to put the fish back in the water. She of course refused. You just can't get good help anymore! We finally released it by scooping it up using a boat paddle and a large dip net.

utes, changing the water as it cools. When boating, the hot exhaust water from an outboard motor can be collected and used for treatment. Chemical hot packs can also be used, and it is a good idea to have a few in your automobile and boat first aid kit.

SKATE AND RAY CONSERVATION

In the Gulf of Mexico the populations of skates and rays appear to be relatively secure. However, there are some species that have been reduced to very low numbers and one that is now endangered and protected by law. The smalltooth sawfish was added to the U.S. Endangered Species List in 1993 and is now protected against capture or harassment. Anyone who catches a smalltooth sawfish *must* release it unmolested and unharmed. Likewise, the largetooth sawfish has been greatly reduced in number and should also be released. The spotted eagle ray, the manta ray, and the devil ray have all experienced reductions in populations and should not be molested. The common species of skates and rays appear to be secure in most areas of the Gulf. However, this does not excuse needless killing of these fish. I have seen fishers on numerous occasions kill stingrays and cast them upon the beach or throw them back into the ocean. I can only assume that the uninformed fisher believes they are doing some service to the marine community or to other fishers. However, a dead stingray either in the water or on land is a greater hazard to beach-goers than a live one, plus no one wants a bunch of decomposing animals on their beach or in the surf. A live stingray will avoid contact with bathers if at all possible and will swim away quickly when disturbed. Therefore, it is best to let the stingrays go unmolested.

Shark Identification

KEY TO THE FAMILIES OF SHARKS OF THE GULF OF MEXICO

(Choose the diagram that best describes your fish.)

1. HEAD SHAPE

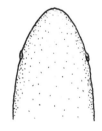

A. Go to #2.

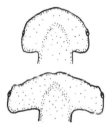

B. Hammerheads. Go to page 79.

C. Angel shark. Go to page 91.

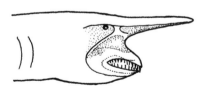

D. Goblin shark. Go to page 118.

2. TAIL SHAPE

A. Go to #3.

B. Thresher sharks. Go to page 93.

C. Great white, mako. Go to page 98.

3. NASAL BARBELS

A. Absent or very small barbels. Go to #4.

B. Prominent nasal barbels present, nurse shark. Go to page 112.

4. DORSAL FIN(S)

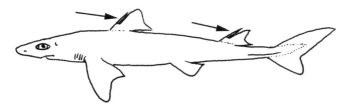

A. Spine(s) preceding one or both dorsal fins. Dogfish sharks. Go to page 103.

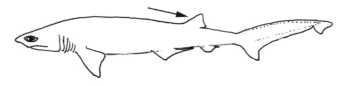

B. A single dorsal fin. Six- or seven-gill sharks. Go to page 87.
C. Two dorsal fins with no spines. Go to #5.

5. FINS/TEETH

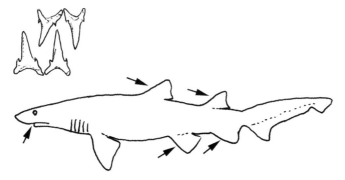

A. Dorsal fins, pelvic fins, and anal fin all similar in size. Slender, blade-like teeth, visible when mouth closed. Sand tiger shark, page 104.

B. Teeth typically triangular in shape, may be serrated. Gray or requiem sharks, page 53.

C. Teeth pavement-like, small, low, and in rows with no sharp cutting edges. Smoothhound sharks, page 109.

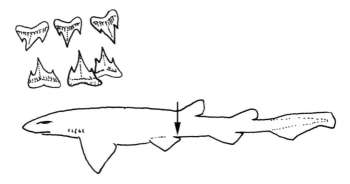

D. Teeth with a triangular middle cusp flanked by a smaller cusp on each side; first dorsal fin over or behind the pelvic fins. Catsharks, page 115.

REQUIEM OR GRAY SHARKS
FAMILY CARCHARHINIDAE

The Gulf of Mexico shark fauna is dominated by members of the family Carcharhinidae, the requiem, gray, or whaler sharks. In this area seventeen species may be encountered, although several rarely enter inshore waters. For example, the blue shark, *Prionace glauca*, and the oceanic whitetip shark, *Carcharhinus longimanus*, are typically found at the surface over deeper waters and rarely venture into shallow inshore areas. The bignose shark, *Carcharhinus altimus*, and the night shark, *Carcharhinus signatus*, are deep-water inhabitants that are found infrequently in shallow waters. Several species within this family are separated from the others by the presence of an **interdorsal ridge**, a prominent ridge between the first and second dorsal fins. This characteristic has prompted these species to be grouped together as the "ridgeback" sharks. The common ridgeback sharks in the family Carcharhinidae include the dusky, silky, and sandbar sharks. The oceanic whitetip shark also has this ridge, but it is not found inshore.

In the shallow inshore waters of the Gulf there are twelve relatively common species. All of the sharks of this family are **viviparous**, meaning they have a **placenta** and umbilical connection between the mother and the developing young. The newborn sharks are called pups and are immediately able to fend for themselves. Litter size ranges

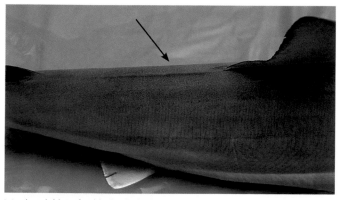
Interdorsal ridge of a ridgeback shark

from as few as one to over one hundred pups. These sharks tend to be benthic—that is, they prefer to be near or on the bottom—and feed predominately on live fish and various crustaceans. Contrary to popular belief, they are not scavengers but are more correctly termed "opportunistic" and prefer living or freshly killed prey. Many of the sharks within this family are very similar in appearance and present a significant problem in identification, especially when juveniles are under scrutiny. The "sand shark" of the northern Gulf is a catchall term for several members of this family. Most of the members of this family pose little or no threat to humans, but a few, owing to their large size, could be dangerous. Economically, there are more requiem sharks harvested by fishers per year than any other group.

KEY TO THE SPECIES OF REQUIEM OR GRAY SHARKS

1. **A.** Dorsal fin broadly rounded with white blotch at tip. Oceanic whitetip shark, page 68.

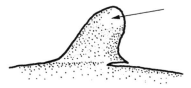

   B. Dorsal fin not broadly rounded, no white blotch. Go to #2.

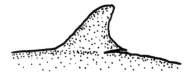

2. **A.** Low fleshy ridge between dorsal fins (interdorsal ridge). Go to #3.

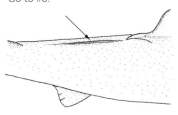

B. No ridge between dorsal fins. Go to #6.

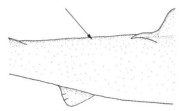

3. **A.** Teeth in upper jaw notched (asymmetrical) and with several large serrations on notched side. Eyes large and green. Nightshark, go to page 73.

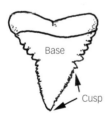

B. Teeth in upper jaw with distinct cusp narrower than tooth base. Base heavily serrated, cusp with finer serrations. Silky shark, go to page 63.

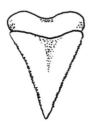

C. Upper jaw teeth broadly triangular in shape. Go to #4.

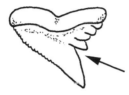

4. A. Dorsal fin height more than half the distance between the first and second dorsal fins (interdorsal distance). Sandbar shark, go to page 71.

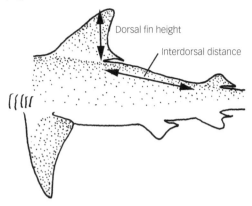

Dorsal fin height

Interdorsal distance

B. Dorsal fin height less than half the distance between the first and second dorsal fins. Go to #5.

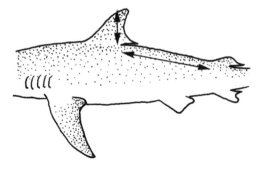

5. A. Pectoral fin and dorsal fin have very concave posterior margins. Pectoral fin relatively slender, its tip more pointed. Dusky shark, go to page 69.

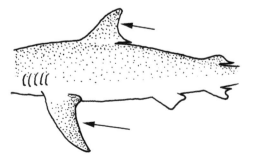

B. Pectoral fin and dorsal fin have slightly concave posterior margins. Pectoral fin tip more rounded. Bignose shark, go to page 61.

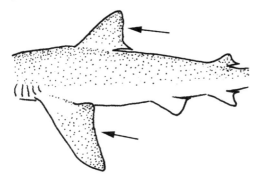

6. **A.** Teeth very oblique, fleshy ridge present on either side of caudal base, small sharks with brown or black spots and stripes on a gray or grayish brown background. Tiger shark, page 74.

Tiger shark upper tooth   Tiger shark fleshy ridge

B. Teeth not strongly notched. Go to #7.

7. **A.** First and second dorsal fins nearly equal in size. Lemon shark, page 76.

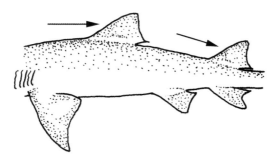

B. Dorsal fins not of equal size. Go to #8.

8. A. Pectoral fin length 3 times longer than dorsal fin height. Blue shark, go to page 77.

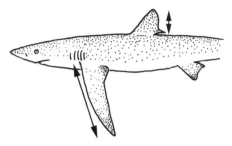

B. Pectoral fin length less than 2.5 times longer than dorsal fin height. Go to #9.

9. A. Snout very short and rounded. Bull shark, go to page 65.

B. Snout not as above. Go to #10.

10. A. Snout with distinct black smudge. Blacknose shark, go to page 60.

B. Snout without a distinct black smudge. Go to #11.

11. A. Elongate creases (labial furrows) at corners of mouth. Atlantic sharpnose shark, go to page 78.

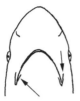

B. No or very short creases at corners of mouth. Go to #12.

12. A. The dorsal fin does not originate over the pectoral fins. Spinner shark, go to page 62.

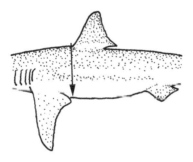

B. The dorsal fin originates over the pectoral fins. Go to #13.

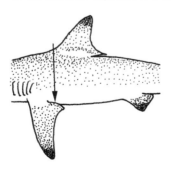

13. A. Fins with no black tips, body color blue to blue-gray when alive or fresh. Finetooth shark, go to page 64.
B. Pectoral, dorsal, pelvic, anal, and lower lobe of caudal fins all tipped in black, body color gray with coppery sheen when alive or fresh. Blacktip shark, go to page 67.

Blacknose shark, *Carcharhinus acronotus*

CHARACTERS: The blacknose shark has a dusky blotch on the tip of the snout. Depending upon season and size, the body color may vary from lemon-yellow, to cream, to gray with a coppery cast. It lacks an interdorsal ridge (see description page 53). Teeth in the upper jaw have narrow, oblique cusps; they are deeply notched laterally and serrated. The lower teeth are slightly oblique, notched laterally, and serrated.

BIOLOGY: The blacknose shark is small; the largest specimens are about 150 centimeters (5 feet) total length. Male blacknose sharks mature at about 100 to 110 centimeters (3.3 to 3.6 feet), and females mature at about 110 to 115 centimeters (3.3 to 3.8 feet). Mating probably occurs in late summer or early fall, and three to six pups are born nine to ten months later in May and June. Pups are born at about 50 centimeters (19 inches) in length. The blacknose shark eats fish, as well as squid, shrimp, and other invertebrates. Although small, care should be exercised when handling this shark owing to their tendency to bite anything within reach.

DISTRIBUTION AND STATUS: The blacknose shark is distributed from North Carolina, around Florida, into the Gulf of Mexico, west to Louisiana, and south to Vitoria and Rio de Janeiro, Brazil. It has been captured out to about 38 meters (125 feet) depth. In the northern Gulf it is consistently captured in shallow waters but is not abundant. In Tampa Bay, Florida, this shark is particularly common around the Sunshine Skyway Bridge. The blacknose is more abundant off the South Carolina coast. The populations of blacknose sharks appear to be secure.

SIMILAR SPECIES: The blacknose shark may be confused with the sharpnose or finetooth sharks. The long labial furrows and white spots in the sharpnose make it easy to identify. The small dusky spot on the tip of the nose in the finetooth shark could cause confusion; however, the smooth, needle-like teeth of the finetooth are very different from the blacknose.

Bignose shark, *Carcharhinus altimus*

CHARACTERS: The snout of the bignose shark is rounded and about as long as or longer than the width of the mouth. An interdorsal ridge (see description page 53) is present. The teeth in the upper jaw are broadly triangular and serrated. The first dorsal fin is large and originates over the inner margins of the pectoral fins. The color varies from gray to bronze on the dorsal surface to whitish below.

BIOLOGY: The bignose shark is an uncommon shark in the Gulf of Mexico preferring deeper waters offshore. It is apparently more common off of south Florida and there is a dubious report of one taken in the northern Gulf of Mexico. This shark feeds on squids, sharks, rays, skates, and other fishes. It is a bottom-dwelling species. Pups are born at about 70 to 90 centimeters (2.3 to 3 feet), and litter size ranges from three to fifteen pups. Maximum size is 282 centimeters (9.3 feet); females mature at 226 centimeters (7.4 feet), and males mature at about 220 centimeters (7.2 feet). This shark could be dangerous because of its size, but the average person would seldom come in contact with this species because it prefers deep water.

DISTRIBUTION AND STATUS: The bignose shark is found in the western Atlantic from Florida to Venezuela and includes the Bahamas, Cuba, Nicaragua, and Costa Rica. It has been captured in depths of 90 to 430 meters (295 to 1,400 feet). There is no information available regarding the status of populations of this shark.

SIMILAR SPECIES: The bignose shark may be confused with the sandbar shark. However, the sandbar shark's snout is shorter and the first dorsal fin is higher.

Spinner shark, *Carcharhinus brevipinna*

CHARACTERS: The spinner shark is characterized by the absence of an interdorsal ridge (see description page 53); a snout that is relatively long and pointed; upper teeth that are narrow, erect, or slightly oblique and uniformly serrated; and lower teeth that are slightly oblique and usually smooth edged. The color is gray to coppery and, in adults, the tips of all the fins may be dusky black. In juveniles, the anal, pelvics, and upper lobe of the caudal may not be tipped with black. The belly is white.

BIOLOGY: The spinner shark is large, up to about 278 centimeters (9 feet) total length. The largest females in the Gulf area are about 230 centimeters (7.5 feet), and the largest males are around 200 centimeters (6.6 feet). Females mature at about 180 centimeters (6 feet), and males at about 170 to 175 centimeters (5.6 to 5.7 feet). Mating appears to be at its peak in June, and after an eleven- to twelve-month gestation, generally six to eight (up to fifteen) pups are born in spring. Pups are born at about 60 to 70 centimeters (2 to 2.3 feet). Spinners eat fish, including small sharks and rays, squids, octopi, and various crustaceans. Spinners are highly active, fast-swimming predators that have the amazing habit of leaping out of the water and spinning. They are particularly prone to do this when hooked.

DISTRIBUTION AND STATUS: The spinner shark is distributed from North Carolina to Florida and the Caribbean and the northern Gulf of Mexico. It is not abundant along the coasts of Mississippi and Alabama but appears to be more common along the northwest Florida coast. The distribution of this species along the Central and South American coasts is not well known because this species is easily confused with the blacktip shark. Because of this confusion, the status of spinner shark populations in the Gulf of Mexico is not known.

SIMILAR SPECIES: In the Gulf of Mexico the spinner is often confused with the blacktip shark. However, the spinner has a longer snout, smaller fins, and a more slender body than the blacktip. The spinner can be readily distinguished from the blacktip by the origin of the first dorsal which is over or slightly behind the inner pectoral fin corner. The more elongate snout of the spinner is also diagnostic. In the blacktip, the origin of the first dorsal is over or just behind the pectoral axil. The black color on the anal fin in the spinner is not a conservative character and cannot be used exclusively to separate the blacktip and the spinner. In the northern Gulf I have observed a number of individuals that have characters intermediate between the spinner and the blacktip, suggesting hybridization between the two.

Silky shark, *Carcharhinus falciformis*

CHARACTERS: The silky shark is one of the ridgeback sharks possessing an interdorsal ridge (see description page 53) between the first and second dorsal fins. The free tip of the second dorsal fin is long, about 1.5 to 3.0 times the height of the second dorsal. The body is slender; the snout is rounded and moderately long, being equal to or slightly less than the width of the mouth. The first dorsal fin is rounded at the top, and it begins behind the free rear tips of the pectoral fins. The teeth in the upper jaw have relatively narrow cusps that are set upon a wide base. The upper portion of the cusp is finely serrated, and the base is coarsely serrated. The teeth in the lower jaw are erect, narrower than the upper teeth, and slightly serrated. The color is dark gray, grayish brown, or bluish black. The belly is grayish or white.

BIOLOGY: The silky is a large shark that grows to a maximum size of about 350 centimeters total length (11.5 feet). Females mature at about 210 to 230 centimeters (7 to 7.5 feet), and males mature at about 200 centimeters (6.5 feet). Little is known of the reproductive biology of this species. A ten- to twelve-month gestation period is likely. Litter size has been reported to be two to eleven pups, and size at birth is about 70 to 85 centimeters (2.3 to 2.8 feet). Silky sharks feed primarily on fishes and squids, but crabs have also been found in their stomachs. This shark has been reported to be dangerous to humans.

DISTRIBUTION AND STATUS: The silky shark is widely distributed in the oceanic waters of the Atlantic. It is typically found in surface waters but has apparently been taken at depths to 500 meters (1,600 feet). In the western Atlantic it has been reported from off Cape Cod in the north to southern Brazil in the south, including the Gulf of Mexico and the Caribbean. The silky is more commonly found in oceanic waters near the continental slope but occasionally strays into inshore waters. In the Gulf of Mexico it is a rather uncommon species in shallow waters. The status of the populations of silky sharks is not known.

SIMILAR SPECIES: The silky can be readily distinguished from other species by its long slender body, characteristic teeth, and long trailing tip of the second dorsal fin.

Finetooth shark, *Carcharhinus isodon*

CHARACTERS: The finetooth shark has no interdorsal ridge (see description page 53) but has a pointed, moderately long snout. The upper and lower teeth are erect, nearly symmetrical, with very narrow, needle-like cusps (the individual blades of the teeth). The upper teeth are irregularly serrated; the base may be moderately serrated or smooth, and the cusps may be serrated or smooth. The lower teeth have no serrations. The finetooth shark is a distinctive gun-metal blue on the back and is white along the lower sides and belly.

BIOLOGY: The finetooth shark is common in inshore waters of the northern and northeastern Gulf of Mexico. In surveys conducted in the northern Gulf it is consistently the second or third most abundant shark captured. Despite its abundance, little is known of the biology of this shark in the Gulf of Mexico. It is a medium-sized shark, reaching a maximum size of about 180 centimeters (6 feet) total length. Females mature at about 137 centimeters (4.5 feet) total length, and males mature at about 120 centimeters (4 feet). Mating most likely occurs in late spring and early summer. In the northern Gulf, newborn finetooth sharks begin to appear in May and are abundant in mid-June. The gestation period is probably eleven to twelve months, and two to six pups are born at an average size of about 58 centimeters (2 feet) total length. Finetooth sharks feed on fishes such as mackerel, whiting, and sea trout. The finetooth shark has a relatively nasty disposition when captured, biting at anything within reach of its jaws. My graduate students and I have been bitten by this shark.

DISTRIBUTION AND STATUS: The finetooth shark is distributed in the western Atlantic from New York to Florida and southern Brazil, including the Gulf of Mexico, Cuba, and the Caribbean. This shark is abundant in the northern Gulf, and its populations appear to be secure.

SIMILAR SPECIES: The finetooth is not easily confused with other sharks owing to its distinctive bluish coloration and the needle-like teeth in both jaws.

Bull shark, *Carcharhinus leucas*

CHARACTERS: The bull shark has a relatively stout body, no interdorsal ridge, and a very short snout that is very rounded and much shorter than the mouth width. The nostrils have a small triangular flap. The upper teeth are broad, triangular, erect to slightly oblique, and serrated. The lower teeth are erect and serrated. The color is typically bluish gray to brownish gray and fading to light gray along the sides. The belly is yellowish white. Pectoral and pelvic fin tips may be dusky black.

BIOLOGY: In the northern Gulf of Mexico, the bull shark is one of the largest sharks commonly found in inshore waters, attaining a total length of about 325 centimeters (10.7 feet). Females mature at about 230 centimeters (7.5 feet) and males at about 220 centimeters (7.2 feet). Size at birth is around 80 centimeters (2.6 feet). In the northern Gulf, bull shark neonates (newborn pups) appear off the coasts of Mississippi and Alabama in summer. A litter size of six is typical, but litters of twelve have been reported. The color is dark gray to brownish gray on the back and white on the belly. The fin tips are dark, particularly in younger individuals. This shark is unique in that it regularly moves into freshwater and, in the United States, has been reported in the Mississippi River as far north as Alton, Illinois. The coastal areas of Louisiana are particularly good bull shark habitat. This shark is large and may be aggressive. Caution should be exercised when encountering large individuals of this species. It has been known to attack humans. Bull sharks feed on mackerels, tuna, other sharks, rays, crabs, shrimp, and will also take carrion (rotting flesh).

DISTRIBUTION AND STATUS: In the western Atlantic, the bull shark is found from New York to Florida, throughout the Gulf of Mexico and the Caribbean, and south to at least Rio de Janeiro, Brazil. This shark has a worldwide distribution in tropical waters. The status of bull shark populations is not known.

SIMILAR SPECIES: The extremely short, rounded snout and triangular upper teeth make it difficult to confuse this shark with others in the area.

THE DOCILE NATURE OF THE BULL SHARK

The image portrayed in the media is that bull sharks are veritable savages. My experiences with hundreds of bull sharks does not fit that image at all. When captured on rod and reel or by net, bull sharks are extremely docile and "good natured." They are absolutely the easiest shark to handle because they typically do not thrash around violently like blacktips, sharpnose, or finetooth sharks. I have seen these other species bite fishing rods, gas tanks, boat wiring, and various human appendages, and one latched onto the vinyl boat seat and would not let go. Bulls, however, will simply lie quietly in the bottom of the boat not making much of a ruckus. This is not meant to suggest that they will not bite because if given the opportunity they may very well do that; they are just not particularly aggressive when captured. Bull sharks seem to "stress out" much less than other sharks when captured, probably because they do not "go berserk." This makes them excellent candidates for biological research and display at public aquaria. This also reinforces my assertion that in most cases when a bull shark bites a bather it is mistaken identity. These sharks are just not ruthless killers.

Blacktip shark, *Carcharhinus limbatus*

CHARACTERS: No interdorsal ridge (see description page 53) is found on the blacktip shark. Its snout is long, about equal to or slightly less than the mouth width, and narrowly rounded to pointed. The top of the dorsal is pointed or narrowly rounded, its origin above or slightly behind the insertion of the pectoral fin base. Upper and lower teeth are similar, both symmetrical with narrow cusps and serrated edges. The color is brownish gray to gray with a coppery sheen in fresh specimens. A dark band extends rearward along each side to about over the origin of the pelvic fin. The fins are tipped in black, being more prominent in juveniles but sometimes fading with age. The pectoral fin tips are always black.

BIOLOGY: In the northern Gulf of Mexico the blacktip is a very common shark, and many are taken by hook-and-line fishers. Female maximum size is about 200 centimeters (6.5 feet) and for males about 180 centimeters (6 feet), although blacktips of 240 centimeters (8 feet) have been reported from other oceans. Females mature at about 160 centimeters (5.2 feet) and six to seven years of age. Males mature at about 135 centimeters (4.4 feet) and four to five years of age. Blacktip sharks mate in late summer and give birth in May and June. Many newborn pups and juveniles can be found in inshore waters during late spring and early summer. The gestation period is about ten months, and one to ten pups are born at about 60 centimeters (2 feet) total length. Blacktips feed on fish, squid, and occasionally crustaceans. Blacktips are fast-swimming, highly active sharks. Like the spinner shark, they are known for leaping from the water and spinning. While not considered to be particularly dangerous, large sharks should be treated with caution.

DISTRIBUTION AND STATUS: Blacktips are distributed worldwide. In the western Atlantic they are found as far north as southern New England and as far south as southern Brazil, including the Gulf of Mexico and the Caribbean. The status of blacktip shark populations in the Gulf of Mexico is not known; however, considering that this species is a prized food fish, it has likely experienced declines over the past few years.

SIMILAR SPECIES: The blacktip and spinner sharks are often confused in the Gulf of Mexico. However, the elongate snout and position of the dorsal fin in the spinner shark reliably separates the two (see spinner shark discussion, page 62).

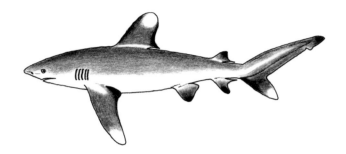

Oceanic whitetip shark, *Carcharhinus maou*

CHARACTERS: An interdorsal ridge (see description page 53) is present, as well as a broadly rounded first dorsal fin, and the first dorsal is noticeably large. The pectorals are long and rounded. The upper teeth are broadly triangular, serrated nearly to the tip, and symmetrical near the front of the mouth but more oblique toward the rear. The lower teeth have broad bases with a slender cusp that are serrated. The color varies from light gray to pale brown or slaty blue above and yellowish to dirty white below. The dorsals are tipped in grayish white, and in some cases the caudal lobes and pectorals are also white tipped.

BIOLOGY: This species is found in oceanic waters in the Gulf of Mexico and rarely, if ever, strays into shallower waters. It prefers tropical waters and does not occur in the western Atlantic when/where water temperatures are lower than about 21°C (70°F). This shark has litters of one to fifteen pups that are born at about 60 to 65 centimeters (about 2 feet) total length. Males mature at 175 to 200 centimeters (5.7 to 6.6 feet) total length, and females at 180 to 200 centimeters (5.9 to 6.6 feet). Maximum size is about 350 to 400 centimeters (11.5 to 13.1 feet). Aggregations of females may be observed on occasions. This shark feeds on fishes, squids, crustaceans, turtles, and decomposing flesh. This shark has probably attacked and eaten human castaways. They are considered dangerous.

DISTRIBUTION AND STATUS: This shark has been recorded from Maine to Argentina, including the Gulf of Mexico. It is an oceanic species rarely observed in shallow waters. This was once an abundant shark in the Gulf of Mexico but is thought to be greatly reduced at this time.

SIMILAR SPECIES: The short, rounded snout, the large, rounded, white-tipped dorsal fin, and the long, rounded white-tipped pectorals make this species easily recognizable from others in the Gulf.

Dusky shark, *Carcharhinus obscurus*

CHARACTERS: The dusky shark has an interdorsal ridge (see description page 53), a short, rounded snout whose length is equal to or less than the mouth width, and a first dorsal fin that is relatively short in height and has a narrowly rounded or pointed tip. The upper teeth are triangluar, erect or moderately oblique, and uniformly serrated. The lower teeth are erect to moderately oblique and serrated. The color is brownish gray to dark gray on the back and white on the belly. The fins, particularly the pectoral, second dorsal, and caudal, are dusky tipped.

BIOLOGY: The dusky is a large shark with the largest specimen reported to be 362 centimeters (12 feet) total length. Females mature at about 289 centimeters (9.4 feet), and males at about 279 centimeters (9.2 feet). Litter size is typically six to ten (average eight), but litters as large as fourteen pups have been recorded. Pups are born at around 85 to 100 centimeters (2.8 to 3.3 feet) total length after a long gestation period, perhaps sixteen months. Little is known of the seasonality of mating or birth in the northern Gulf of Mexico. The dusky feeds on bottom fishes primarily, including other sharks. Large dusky sharks should be treated with caution.

DISTRIBUTION AND STATUS: The dusky is found from Georges Bank south to Florida including the Gulf of Mexico, parts of the Caribbean, and Brazil. In the northern Gulf of Mexico, the dusky shark was at one time common. The author has observed very large duskys landed in northern Gulf of Mexico fishing tournaments in the early 1980s. The area around Pensacola, Florida, was even referred to as the "land of the big dusky." Unfortunately, this shark has virtually disappeared from the area. The apparently critical status of dusky shark populations in the northern Gulf of Mexico suggests that it should be added to the list of prohibited species in both state and federal waters.

SIMILAR SPECIES: Dusky sharks may be confused with sandbar sharks, but the sandbar shark has a very high dorsal fin that is placed further forward on the body. The sandbar has a more robust appearance than the dusky. The bull shark has a much shorter snout, and the silky has upper teeth possessing heavily serrated bases.

Caribbean reef shark, *Carcharhinus perezi*

CHARACTERS: The Caribbean reef shark has an interdorsal ridge (see description page 53), a snout that is short and blunt, teeth with serrated edges, upper jaw teeth that are symmetrical near the front of the mouth but more oblique toward the rear, lower jaw teeth that are all nearly erect, and all teeth have a broad base with a relatively narrow cusp. The first dorsal fin is relatively small with a short rear tip. This shark is grayish brown or grayish olive above and the belly is white or yellowish olive.

BIOLOGY: The Caribbean reef shark is found over coral reefs and is therefore more common in the southern Gulf of Mexico. It is one of the most abundant sharks around the Bahamas and the Antilles. It is a sluggish bottom-dwelling species and may occasionally rest on the bottom. Its maximum known size is about 300 centimeters (9.8 feet); males mature at about 150 to 170 centimeters (4.9 to 5.6 feet) total length; females mature at about 200 to 295 centimeters (6.6 to 9.7 feet); and size at birth is around 70 centimeters (2.3 feet). Litter size is four to six pups. This species feeds on bony fishes. This is a dangerous species and has been reported to attack humans.

DISTRIBUTION AND STATUS: The Caribbean reef shark ranges from Bermuda to Florida and the Gulf of Mexico, the Bahamas, and the Antilles to southern Brazil. It is usually found at less than 30 meters (about 100 feet) depth. It is not common in the Gulf of Mexico as it is a coral reef species. It has apparently been reported from the northwestern Gulf. Little is known of the population status of the species.

SIMILAR SPECIES: This shark could be confused with a number of similar species. The bignose shark has a longer snout and the upper teeth are broadly triangular. The dusky shark also has a longer snout, more triangular teeth, and the length of the posterior lobe of the second dorsal fin is about twice the height of that fin. The silky shark's dorsal fin originates behind the rear edge of the pectoral fin whereas the present species dorsal originates directly over or slightly ahead of the pectoral rear edge. The sandbar shark has a much larger dorsal that originates well ahead of the rear edge of the pectoral fin.

Sandbar shark, *Carcharhinus plumbeus*

CHARACTERS: The sandbar shark has an interdorsal ridge (see description page 53). The body has a very robust appearance and the first dorsal fin is very high, set far forward on the body, and triangular with a pointed or narrowly rounded tip. The origin of the first dorsal is over or anterior to the axil (where the trailing edge of the fin attaches to the body) of the pectoral. The teeth in the upper jaw are finely serrated, triangular, and erect to slightly oblique. The lower teeth have narrow erect cusps and are also finely serrated. The color is gray to brownish and the belly is whitish.

BIOLOGY: This is a large shark growing to a total length of about 300 centimeters (10 feet). Females and males mature at about 180 to 190 centimeters (6 to 6.2 feet) total length. Mating occurs in spring and summer, with litter sizes ranging from one to fourteen with an average of nine pups. Size at birth is about 60 centimeters (2 feet). In the northern Gulf of Mexico an important nursery area exists around Cape San Blas, Florida. Juvenile sandbar sharks are occasionally captured off of Mississippi and Alabama, suggesting that other nursery grounds may exist. The sandbar shark is a coastal species but has been reported to depths of 200 meters (650 feet). The sandbar shark feeds on flatfish, stingrays, schooling fishes, squid, and crabs. Sandbar sharks tagged off of the East Coast of the United States have been recaptured in the Gulf, suggesting an extensive migration around the tip of Florida.

DISTRIBUTION AND STATUS: The sandbar shark is distributed worldwide, and in the western North Atlantic it occurs from Cape Cod, Massachusetts, to Brazil, including the Gulf of Mexico and the Caribbean. The status of the sandbar shark in the Western Atlantic and the Gulf of Mexico may be in question. Sandbar shark commercial landings in U.S. southern waters account for 60 percent of the total catch. The decline in shark landings observed over the last several years suggests that sandbar populations are likely declining as well.

SIMILAR SPECIES: The sandbar shark could be confused with the dusky or silky, particularly as juveniles. However, the forward position of the first dorsal fin and the short free margin of the second dorsal separates it from the dusky and the silky, respectively.

Smalltail shark, *Carcharhinus porosus*

CHARACTERS: The smalltail shark lacks an interdorsal ridge (see description page 53) and has a snout that is usually longer than the width of the mouth; the first dorsal fin is low and has a narrowly rounded tip, and the origin of the second dorsal fin is over or slightly behind the midpoint of the anal fin. The teeth in the upper jaw are serrated; the cusps are erect in the front of the jaw and become more oblique toward the sides. The teeth in the lower jaw have high, narrow cusps that are erect in front and oblique posteriorly. Color is grayish and the belly and sides are white. The sides and pelvics are reddish in some individuals.

BIOLOGY: Little is known of the biology of the smalltail shark. The smalltail reaches a maximum size of about 120 centimeters (4 feet) total length. Maturity in females is reached at about 84 centimeters (2.8 feet) and in males at about 75 centimeters (2.5 feet). Litter size ranges from two to seven, with an average of five, and the pups are born in the spring at 31 to 40 centimeters (12 to 16 inches). The smalltail feeds on fishes and small invertebrates, principally crabs. The smalltail prefers mud bottoms near the mouths of large rivers. This species is harmless.

DISTRIBUTION AND STATUS: The distribution of the smalltail is not well known. It has been captured from off Mississippi, Louisiana, and Texas and is found along the Gulf coast of northern Mexico. It has not been recorded from much of the Caribbean coast of Central America but is found from Venezuela to southern Brazil. The status of smalltail populations is not known. This species is apparently not found east of the Mississippi coast.

SIMILAR SPECIES: The origin of the second dorsal of the smalltail is behind the origin of the anal. The only other shark with this characteristic in our area is the Atlantic sharpnose shark. The sharpnose shark has long labial folds at the corners of the mouth whereas the smalltail labial folds are much shorter.

Nightshark, *Carcharhinus signatus*

CHARACTERS: The nightshark has an interdorsal ridge (see description page 53). It has a relatively small first dorsal fin with a narrowly rounded apex. The body is slim, the snout is elongate, narrow, and relatively pointed. The teeth are smooth edged or with weak serrations. The upper teeth are oblique, their bases having two to several cusplets or strong serrations. Lower teeth are nearly erect with a slender cusp. The night shark is grayish blue with scattered black spots above and grayish white below. This shark's most distinguishing characteristic is its large green eyes.

BIOLOGY: This is a poorly known species. Maximum size is about 280 centimeters (9 feet), size at birth is about 60 centimeters; litter sizes are reported to be four to twelve pups. This shark preys on fishes and squids. Off Cuba it is said to be nocturnal. It is not known to be dangerous to humans.

DISTRIBUTION AND STATUS: The nightshark is found from Delaware to Florida and the Gulf of Mexico through the Antilles to Argentina. This is a deep-water species found below about 100 meters (330 feet) depth in the Gulf, although in some parts of the world it has been recorded from very shallow water. No information is available concerning population status.

SIMILAR SPECIES: The large green eyes, the mid-dorsal ridge, and the long pointed snout make this shark easily identifiable.

Tiger shark, *Galeocerdo cuvieri*

CHARACTERS: The tiger shark has a blunt snout that is much shorter than the width of the mouth. There are small openings called spiracles present behind each eye. The teeth possess very coarse serrations that are themselves serrated; the teeth are deeply notched, and the tips are very oblique. There is a low keel on either side of the caudal peduncle. When young, the tiger shark is strikingly colored having black spots and bars on a gray or grayish-brown background. With growth the spots coalesce into stripes along the sides, but these may eventually become very faint in the largest individuals. The belly is white to yellowish.

BIOLOGY: The tiger shark is one of the largest sharks in our area. While adults are typically 350 to 400 centimeters (11.5 to 13 feet) total length, in some areas they may attain a maximum length of at least 650 centimeters (21 feet). Females mature at about 300 centimeters (10 feet) and males at about 270 centimeters (9 feet). Unlike the other sharks in this family, tiger shark development is by aplacental viviparity; that is, there is no placental connection between the developing embryo and the mother shark. Tiger sharks produce large litters with eighty-two embryos reported in one case. The pups are born at 68 to 77 centimeters in spring. The gestation period is believed to be about twelve months. Tigers prey upon sea turtles, seabirds, dolphins, rays, sharks, schooling fishes, crabs, conchs, horseshoe crabs, and a host of other items. The tiger shark's habit of ingesting inanimate objects (e.g., a roll of roofing felt, lumps of coal, tin cans, aluminum foil) has helped to perpetuate the myth that sharks are mindless eating machines. However, some scientists have suggested that tiger sharks may consume some of these items to help regulate buoyancy, the same way that a submarine takes on or drops ballast to descend or ascend in the water. Although these sharks may frequent shallow waters, they are not often encountered inshore and have also been collected far offshore. Given the tiger's large size and apparently undiscriminating palate, it is a shark that should be treated with respect. This shark has been implicated in attacks on humans.

DISTRIBUTION AND STATUS: Along the U.S. Atlantic coast, this shark occasionally strays as far north as Rhode Island. It is commonly taken in North and South Carolina waters and is found around the tip of Florida, throughout the Gulf of Mexico and the Caribbean Sea. Southward this species can be found to northern Argentina. Tiger shark populations have likely declined over the past several years, and large individuals are now seen much less frequently in the northern Gulf of Mexico.

SIMILAR SPECIES: The tiger shark is not likely to be confused with any other species owing to its distinctive coloration, particularly when young, and its extremely short, blunt snout.

TIGER SHARK GLUTTONY

As a graduate student out of the Dauphin Island Sea Lab, I spent many hours longlining (hundreds of hooks on miles of line) aboard the research vessel *Rounsefel* in the late 1970s and early 1980s. On several trips I was very seasick and threw up repeatedly into the blue waters of the Gulf. As a matter of fact, I set for myself a goal of throwing up in all of the major oceans of the world. Unfortunately, my problem with motion sickness has long since abated but not before "christening" the Gulf of Mexico, the Caribbean Sea, and the Atlantic Ocean. On one trip, made memorable because of the intensity of my personal "chumming," I experienced first-hand the gluttony of the tiger shark. We were about fifty miles south of Mobile Bay and were capturing a good many sharks including an extremely large tiger. It took all hands and a winch to get the shark on board. It was huge, perhaps six hundred pounds, and we assumed that its large girth was because it was pregnant. One of my fellow graduate students performed an autopsy and, instead of shark pups, out spilled an incredible number of fish in various stages of digestion. There was literally a mountain of fish on the deck of the vessel. At least somebody had an appetite on that trip. The tiger must have been following a shrimp trawler and had consumed hundreds of fish that were discarded by the shrimpers. Sharks are known to follow trawlers waiting for net haul-up. The sound of the winches on haul-up is like ringing a dinner bell for sharks. The amazing thing about this tiger shark was that despite the fact that it was engorged with fish, it still took a baited hook. Its gluttony led to its demise!

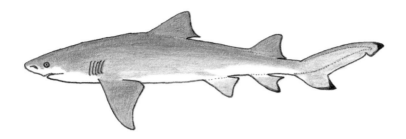

Lemon shark, *Negaprion brevirostris*

CHARACTERS: The lemon shark has no interdorsal ridge, a broad short snout that is shorter than the width of the mouth, first and second dorsal fins that are similarly sized, and a first dorsal whose origin is behind the free rear tips of the pectoral fin bases. The teeth have narrow, erect, smooth-edged cusps becoming more oblique toward the rear of the jaw. The bases of the upper teeth are serrated. The color is yellowish brown, dark brown, or olive gray. The belly is whitish or yellowish.

BIOLOGY: The lemon shark is a large shark attaining a maximum length of about 320 centimeters (10.5 feet). Females mature at about 240 centimeters (8 feet) and males at about 225 centimeters (7.4 feet) total length. Mating occurs in the spring, and pups are born in spring and summer after a twelve-month gestation. Litter size is typically five to seventeen embryos. Pups are born at about 60 to 65 centimeters (2 to 2.1 feet) total length. Lemon sharks are sluggish, bottom-oriented sharks, that feed on schooling fish, catfish, ray, shrimp, crab, squid, and octopus. They are common inhabitants of shallow, mangrove swamps. This is a large shark that should be considered dangerous.

DISTRIBUTION AND STATUS: The lemon shark is found from North Carolina to northern Brazil including the Gulf of Mexico and the Caribbean Sea. In the northern Gulf of Mexico, this shark is more common east of Mobile Bay. There is little information about the status of populations of this species.

SIMILAR SPECIES: The similarity in size of the first and second dorsal fins, the coloration, the unique teeth, and the short, broad snout readily distinguishes this species from other sharks.

Blue shark, *Prionace glauca*

CHARACTERS: The blue shark has no interdorsal ridge. The body is very slender and bullet-shaped. The snout is very long and slender. The first dorsal fin originates well behind the pectoral fin rear margins. The pectoral fins are very long and narrow. There is a weak keel along the sides of the body where the caudal fin attaches to the body. The teeth in the upper jaw are serrated, triangular, and broadly curved. The lower jaw teeth are more slender and are serrated. The blue shark is dark blue above, bright blue along the sides, and white below. The pectoral and anal fin tips have dark smudges.

BIOLOGY: The blue shark is an oceanic species rarely venturing near shore. Maximum reported size is about 380 centimeters (12.5 feet), males mature at about 180 to 280 centimeters (6 to 9 feet) and females at 220 to 320 centimeters (7.2 to 10.5 feet). The litter size can be very large—up to 135 pups. Pups are born at 35 to 44 centimeters (1.1 to 1.4 feet). This shark feeds on fishes, other sharks, seabirds, squids, and decomposing animals. It is considered dangerous to humans.

DISTRIBUTION AND STATUS: The blue shark is the most widely distributed of all sharks and is found in all tropical and warm temperate seas. It prefers the shallow, sunlit, open-ocean habitat. It ranges from Newfoundland to Argentina including, although rare, the Gulf of Mexico.

SIMILAR SPECIES: The bright blue coloration, dorsal fin set far back on the body, and long, slender snout and body make this shark very distinctive and difficult to confuse with other species.

Atlantic sharpnose shark, *Rhizoprionodon terraenovae*

CHARACTERS: The sharpnose shark is readily identified by its coloration; gray to grayish often with white spots along the dorsal surface. The fins are edged in black. The long labial furrows (the folds of skin at the corners of the mouth) are also diagnostic. The teeth are unserrated and are asymmetrical in both jaws.

BIOLOGY: The Atlantic sharpnose is the most common shark found in the northern Gulf of Mexico. It is small, with maximum size about 107 centimeters (3.5 feet) total length. Males mature at about 80 centimeters (2.6 feet) and females at about 85 centimeters (2.8 feet). The peak in mating activity occurs between mid-June and mid-July. The gestation period in the sharpnose shark is ten to eleven months. Outside of a few short months in summer, it is very unusual to find an adult female that is not gravid. It is likewise unusual to find adult females in shallow coastal waters. Most adult females are found just offshore in deeper waters. Sex and size segregation is commonly seen in sharpnose populations. The litter size ranges from one to seven pups, but in most cases either four or six will be present. Pups are born in May or June at an average size of about 33 centimeters (13 inches) total length. The shallows of the northern Gulf of Mexico's extensive barrier island system serve as important pupping/nursery grounds for these sharks. Our present best estimates suggest that sharpnose sharks rarely live to be greater than ten years old. Sharpnose sharks prey upon fish primarily but will also take squid, shrimp, mantis shrimp, and other invertebrate prey.

DISTRIBUTION AND STATUS: The sharpnose shark is distributed from the Bay of Fundy to the Yucatan Peninsula. Seasonally, sharpnose sharks engage in primarily an inshore-offshore movement, leaving the coast for warmer offshore waters during October to November and returning again in March to April. Sharpnose shark populations are apparently secure throughout their range. However, population analyses suggest that improved fisheries and biological data are needed to reassess the status of this shark.

SIMILAR SPECIES: The sharpnose shark may be confused with the smalltail shark. However, the smalltail does not have white spots and the sharpnose has long labial folds at the corners of the mouth. The smalltail seems to be more common in the western Gulf of Mexico.

HAMMERHEAD SHARKS
FAMILY SPHYRNIDAE

Of all the sharks in our area the hammerhead sharks are arguably the oddest of the lot. The winglike lateral expansions of the head give the impression that these sharks were designed by a deranged aerospace engineer. The expanded head has been variously reported as important in increased olfactory ability, enhanced navigational ability, improved prey location, and increased maneuverability. The truth is probably a combination of several, if not all, of the above. The hammerhead sharks are very bottom oriented. When the stomach contents of these sharks is examined, primarily bottom-dwelling organisms such as stingrays, crabs, shrimps, and lobsters are found. It is not too difficult to imagine a hammerhead shark swimming just above the ocean's floor, using its expanded head like a metal detector to find buried food items. The scalloped and great hammerhead sharks seem to be particularly fond of preying upon stingrays. In one case, I found over thirty stingray spines in the stomach of an adult great hammerhead. Additionally, there were several spines stuck in the gums and even in the jaws of the shark.

In decreasing order of abundance, the bonnethead, scalloped, and great hammerhead sharks are the most frequently encountered in the Gulf of Mexico. Identification of the species of hammerheads is based upon the shape of the head. The easiest hammerhead shark to identify using head shape is the bonnethead shark. The head is less expanded than the other hammerheads and has the appearance of a lady's bonnet or a shovel. The bonnethead shark is sometimes called the shovelnose shark for obvious reasons. The hammer of the scalloped hammerhead has a more swept-back appearance and the front edge has a prominent notch in its center. The head of the great hammerhead shark is nearly straight across its front margin, and there is no notch. The other hammerhead sharks that may be encountered in the Gulf are the smalleye and the scoophead, although these species are possibly found only in the extreme southern Gulf of Mexico.

KEY TO THE SPECIES OF
HAMMERHEAD SHARKS

1. A. The hammer is bonnet or spade shaped. Bonnethead shark.
Go to page 84.

B. The hammer is not bonnet or spade shaped. Go to #2.

2. A. There is no notch in the center of the front edge of the head.
Smooth hammerhead shark. Go to page 86.

B. There is a notch in the front edge of the head. Go to #3.

3. A. The front edge of the hammer is relatively straight.
Great hammerhead shark. Go to page 83.

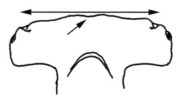

B. The front edge of the head is not straight. Go to #4.

4. **A.** The center of the eye is about on the same line with the front of the mouth. Scalloped hammerhead shark. Go to page 82.

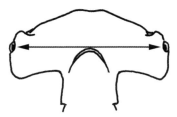

B. The center of the eye is not on the same line with, but is well ahead of, the front of the mouth. Smalleye hammerhead shark. Go to page 85.

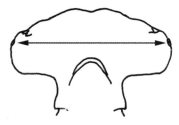

Scalloped hammerhead shark, *Sphyrna lewini*

CHARACTERS: The scalloped hammerhead is olive, to gray, to brownish gray above, but shades to white below. The head is laterally expanded and has a swept-back appearance. There is a distinct median notch on the anterior margin of the head. The first dorsal fin is very high. The teeth are triangular, usually with smooth-edged cusps or slightly serrated in large individuals.

BIOLOGY: The scalloped hammerhead is not an uncommon shark in the Gulf of Mexico but is indeed the most common large hammerhead encountered in this area. The maximum size is about 420 centimeters (14 feet), but individuals about 360 centimeters (12 feet) are common. Females mature at about 210 centimeters (7 feet) and males at about 150 centimeters (5 feet) total length. The gestation period is about ten months, and litter size ranges from fifteen to thirty-one pups. Pups are born in late spring and summer at 40 to 50 centimeters (16 to 20 inches) total length. These sharks feed on squids, octopi, lobsters, shrimps, crabs, and bottom fishes, including other sharks, and are efficient predators of stingrays.

DISTRIBUTION AND STATUS: The scalloped hammerhead is distributed from New Jersey, throughout the Gulf of Mexico, and south to Brazil. The status of scalloped hammerhead populations is not known, but considering the fact that they occur in heavily fished areas, and are frequently taken by fishermen, their numbers are probably reduced.

SIMILAR SPECIES: The scalloped hammerhead could be confused with the great hammerhead shark, and in fact, I have heard fishers call large scalloped hammerhead sharks "great hammerheads." The straight front margin of the head of the great hammerhead and the deeper notch in the center of the front margin in the scalloped reliably separates the two.

Great hammerhead shark, *Sphyrna mokarran*

CHARACTERS: The great hammerhead is light gray to brownish gray above but shades to white below. The leading edge of the head is very straight. There is a much shallower median notch on the leading edge of the head. The first dorsal fin is very high. The teeth are triangular and strongly serrated.

BIOLOGY: The great hammerhead is not common in inshore waters in the Gulf of Mexico. Great hammerheads are more common over deeper, offshore waters although they certainly venture into shallower waters. The maximum size is about 600 centimeters (19.7 feet) but the largest are typically about 400 centimeters (13 feet). Females mature at about 300 centimeters (10 feet). The gestation period is estimated to be about ten months, and litter size ranges from thirteen to forty-two pups. Pups are born in summer at about 56 to 70 centimeters (22 to 28 inches) total length. These sharks are specialized for feeding on stingrays but also take fishes, sea catfishes, crabs, squids, other sharks, and lobsters. Because of its large size and the fact that this shark has been implicated in attacks on humans, it should be treated with caution.

DISTRIBUTION AND STATUS: The great hammerhead is distributed from North Carolina, throughout the Gulf of Mexico, south to Uruguay. The great hammerhead tends to be found more commonly in tropical waters. The status of great hammerhead populations is not known, but owing to its very large size (most of the large shark species are drastically reduced in number) and the fact that they are taken as bycatch in several fisheries, their populations are probably declining.

SIMILAR SPECIES: The great hammerhead could be confused with the scalloped hammerhead shark. See description of similar species for the scalloped hammerhead.

Bonnethead shark, *Sphyrna tiburo*

CHARACTERS: The bonnethead is grayish to olive in color with a scattering of small black specks on the dorsal surface. The head is more rounded than other hammerhead sharks, and there is no median notch on the anterior margin of the head. The teeth in the upper jaw are unserrated, and have oblique cusps. The lower jaw teeth have symmetrical, unserrated cusps in the front of the jaw, changing to pavement-like, crushing teeth in the rear.

BIOLOGY: The bonnethead shark is a common shark in the Gulf of Mexico, but in the northern Gulf it is more common east of Mobile Bay, particularly in the shallow waters of northwest Florida and further south. The bonnethead is small, with a maximum size of about 109 centimeters (3.6 feet) total length for males and 124 centimeters (4.1 feet) for females. The average size is about 80 centimeters (2.6 feet) for males and 98 centimeters (3.2 feet) for females. Males mature at about two years of age and females at about two and a half years. Bonnethead sharks apparently mate during fall, and the females store sperm until the following spring when the eggs ovulate and are fertilized. Gestation is four to five months, the shortest of any placental viviparous shark species. Maximum litter size in this shark is about sixteen pups, but most females carry an average of about ten pups. Pups are born in August to September at about 35 centimeters total length (1.1 feet). Maximum age in bonnethead sharks is about eight years for males and twelve years for females. Bonnethead sharks are bottom oriented and feed almost exclusively on blue crabs but will also take shrimp, squid, and the occasional fish.

DISTRIBUTION AND STATUS: The bonnethead shark is distributed from New England, throughout the Gulf of Mexico, and south to Brazil. It is the dominant inshore species along the Gulf coast of Florida. Like most of the sharks of the northern Gulf of Mexico, they engage in an inshore/offshore migration, appearing in shallow waters in the spring and departing in the fall. Their populations seem to be secure.

SIMILAR SPECIES: The bonnethead could not be easily confused with any other hammerhead shark owing to the rounded, shovel shape of the head. The other hammerheads in this area are significantly larger and have a much greater expansion of the head.

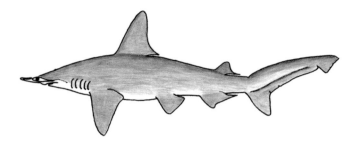

Smalleye hammerhead shark, *Sphyrna tudes*

CHARACTERS: The smalleye hammerhead is grayish brown above and is the same color but lighter below. The eye is relatively small. A deep, conspicuous notch is found in the middle of the front margin of the head. The first dorsal fin is moderately high. The teeth are smooth edged. The upper teeth are asymmetrical, tending to point rearward, while the lowers are more slender and erect.

BIOLOGY: The smalleye hammerhead is uncommon in the Gulf of Mexico and records of this species in the Gulf may be misidentifications. The maximum size is about 150 centimeters (5 feet). Males mature at about 120 centimeters (4 feet) and females at about 130 centimeters (4.3 feet). Litters of six to nine pups have been observed, and size at birth is about 30 centimeters (1 foot) total length. These sharks feed on squids, shrimps, crabs, and bottom fishes, including other sharks. They likely feed on stingrays and skates.

DISTRIBUTION AND STATUS: The smalleye hammerhead has been reported from the northern Gulf of Mexico and from Venezuela and Uruguay. They are found out to about 12 meters (40 feet). The status of smalleye hammerhead populations in the Gulf is not known because it is a very uncommon species.

SIMILAR SPECIES: The smalleye hammerhead could be easily confused with other hammerhead sharks that possess a notch in the middle of the front margin of the head, particularly the scalloped hammerhead. However, the very small eye and its position well ahead of the front of the mouth separates it from others.

Smooth hammerhead shark, *Sphyrna zygaena*

CHARACTERS: The smooth hammerhead is olive to brownish gray above but shades to white or grayish white below. The pectorals are black tipped in some. The head has a smooth front margin with no notch in the middle of the margin. The first dorsal fin is very high. The teeth in the upper jaw are smooth edged in small specimens but have weak serrations in larger sharks. The front teeth tend to be symmetrical and erect, but the others are very strongly notched and asymmetrical.

BIOLOGY: The smooth hammerhead has been recorded from the southern Gulf of Mexico off Florida, but it is very uncommon. The maximum size is about 396 centimeters (13 feet). Adults mature at about 180 to 213 centimeters (6 to 7 feet). The gestation period is about ten months, and litters of about thirty to forty pups have been observed. Pups are born at about 50 centimeters (20 inches) total length. These sharks feed on squids, octopi, lobsters, shrimps, crabs, and bottom fishes, including other sharks, stingrays, and skates.

DISTRIBUTION AND STATUS: The smooth hammerhead has been reported from Massachusetts to south Florida, the southern Gulf of Mexico, and south to Uruguay and northern Argentina. The status of smooth hammerhead populations in the Gulf is not known because it is a very uncommon species.

SIMILAR SPECIES: The smooth hammerhead could not be easily confused with any other hammerhead because of the absence of a notch in the middle of the front margin of the head. Only the bonnethead shark shares this characteristic, and it has a very different head shape.

SIX- AND SEVEN-GILL SHARKS
FAMILY HEXANCHIDAE

Most sharks have only five gill slits, but this family contains those with six or seven. The gill slits are very long, extending under the head of the shark. The eyes are large, there is a single dorsal fin, and the lower lobe of the caudal is not well developed. A distinctive characteristic of this group is the broad comblike teeth in the lower jaw, completely different from the upper jaw teeth. The six- and seven-gill sharks are more common offshore, in the deeper waters of the Gulf of Mexico; however, in some parts of the world, particularly northern waters, they can be found at very shallow depths. These sharks are bottom-dwelling species, and the six-gill has been described as very sluggish. There are three species recorded from the Gulf of Mexico.

KEY TO THE SPECIES OF
SIX- AND SEVEN-GILL SHARKS

1 **A.** There are seven gill slits. Seven-gill shark. Go to page 88.
   B. There are six gill slits. Go to #2.

2. **A.** The snout is short and blunt; the lower jaw has 6 rows of large comblike teeth on each side of the jaw. Distance A. and B. in diagram are about equal. Six-gill shark. Go to page 89.

B. The snout is longer and more pointed, the lower jaw has 5 rows of large comblike teeth on each side of the jaw. Distance A. is much less than distance B. Bigeye six-gill shark. Go to page 90.

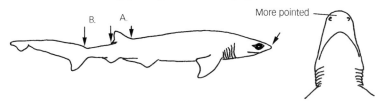

Seven-gill shark, *Heptranchias perlo*

CHARACTERS: This shark is elongate in shape; the snout is pointed; the eyes are large; the seven gill slits are long; the lower lobe of the caudal is poorly developed; the teeth in the lower jaw are comblike, with the second or third cusplet being long and the others smaller. The color is uniformly gray with shades of brown, lighter in color below. The pectorals have white edges, the dorsal is black at its tip with two white spots.

BIOLOGY: This shark is not particularly common in the Gulf of Mexico, and its biology is poorly known. It is a bottom-dwelling species. Maximum size is about 140 centimeters (4.6 feet); males mature at about 85 centimeters (2.8 feet), and females at about 90 centimeters (3 feet) total length. Size at birth is about 26 centimeters (10 inches). Litter size is nine to twenty pups. It feeds on fishes, squids, and most likely a wide variety of bottom organisms. This shark is relatively harmless.

DISTRIBUTION AND STATUS: The seven-gill shark is a bottom-dwelling species found between 27 and 1,720 meters (90 and 560 feet) depth. In the western Atlantic it is found in the Gulf and in the Caribbean. The status of this species is not known.

SIMILAR SPECIES: The seven gill slits separate this species from any other shark in the Gulf.

Six-gill shark, *Hexanchus griseus*

CHARACTERS: The six-gill shark is broader than the other species in this family; the snout is short and blunt; and the head is wide and flattened. The eyes are large and oval; the six gill slits are long; the lower lobe of the caudal is poorly developed; the teeth in the lower jaw are comblike, with the second or third cusplet being longest and the others progressively smaller. There are six rows of teeth on each side of the lower jaw. The color is described as coffee or very dark gray above and the same color but lighter below.

BIOLOGY: This shark is not particularly common in the Gulf of Mexico, and its biology is poorly known. It is a bottom-dwelling species. Maximum size is about 140 centimeters (4.6 feet); males mature at about 85 centimeters (2.8 feet) and females at about 90 centimeters (3 feet) total length. Size at birth is about 26 centimeters (10 inches). Litter size is nine to twenty pups. It is believed to mate in the spring and autumn. Six-gills feed on fishes, crabs, shrimps, and parts bitten off of other sharks that were hooked. They will come to the surface to take discarded fish. This shark may make a migration from deeper waters to the surface at night to feed. This shark is too small and uncommon to be considered a threat to humans.

DISTRIBUTION AND STATUS: The six-gill shark is a bottom-dwelling species found from the surface to about 1,875 meters (6,200 feet) depth. In the western Atlantic it has been captured off of North Carolina and is found in the Gulf and in the Caribbean. The status of this species is not known.

SIMILAR SPECIES: The six rows of teeth in the lower jaw separate this species from the closely related bigeye six-gill shark, which has five rows.

Bigeye six-gill shark, *Hexanchus vitulus*

CHARACTERS: This shark is more slender bodied than the six-gill, and the snout is short and more pointed. The head is also not as wide as the six-gill. The eyes are large; the six gill slits are long; the lower lobe of the caudal is poorly developed; the teeth in the lower jaw are comblike, with the cusplets progressively smaller laterally. There are five rows of teeth on each side of the lower jaw. The color is described as dark gray above and lighter below.

BIOLOGY: This shark is not particularly common in the Gulf of Mexico, and its biology is poorly known. It is a bottom-dwelling species. Maximum size is about 180 centimeters (6 feet); males mature at 120 to 160 centimeters (4 to 5.2 feet) and females at 140 to 180 centimeters (4.6 to 6 feet) total length. Size at birth is about 40 centimeters (16 inches). Litter size is about thirteen pups. The bigeye six-gill feeds on fishes, crabs, shrimps, and other bottom organisms. This shark is too uncommon to be considered a threat to humans.

DISTRIBUTION AND STATUS: The bigeye six-gill shark is a bottom-dwelling species found between about 90 and 600 meters (300 and 1,970 feet) depth. In the western Atlantic it has been reported from the Bahamas, south Florida, the Gulf of Mexico, Nicaragua, and Costa Rica. The status of this species is not known.

SIMILAR SPECIES: See description of the six-gill shark.

ANGEL SHARKS
FAMILY SQUATINIDAE

Although there are a number of different kinds of angel sharks worldwide, we have only a single species in the Gulf of Mexico. The angel shark is one of the most easily recognized sharks in the Gulf, and because there is only one species here, it cannot be confused with any other shark. It looks like a cross between a shark and a stingray because of its very large pectoral fins. As a matter of fact, some scientists believe the angel sharks may be a sort of "missing link" between the sharks and the skates and stingrays. However, while the stingrays and skates have their pectoral fins attached to the head, the angel shark does not.

Angel shark, *Squatina dumeril*

CHARACTERS: This shark is very flattened and raylike; the head has a distinct neck; and the expanded pectoral fins are not attached to the head. The name "angel" comes from the expanded pectorals appearing winglike. The dorsal fins are small and positioned far back on the tail, and the pelvic fins are expanded. The caudal fin is poorly developed, and there is no anal fin. The teeth are similar in both jaws and appear as a single, conical, unserrated cusp on a broad base. The mouth is terminal, and the gill openings are not visible from above. The color is bluish gray or ashy gray above, with red on the head and margins of fins and white below. There may be irregular red marking on top and bottom.

BIOLOGY: This shark is not particularly common in the Gulf of Mexico and is poorly known. Maximum size is about 150 centimeters (4.9 feet); males and females are mature at 92 to 107 centimeters (3 to 3.5 feet). It apparently buries itself in sand or mud and ambushes prey such as bottom fishes, squids, octopi, and crabs. Litter size is up to sixteen pups that are probably born in summer. This shark is relatively harmless but it can inflict a nasty bite.

DISTRIBUTION AND STATUS: The angel shark is a bottom-dwelling species that has been captured in very shallow water but also at a depth of 1,300 meters (4,265 feet). This shark is found from Massachusetts to the Florida Keys, including the Gulf of Mexico, and as far south as Venezuela. The status of this species is not known.

SIMILAR SPECIES: This is a very characteristic species and could not be confused with any other.

THRESHER SHARKS
FAMILY ALOPIIDAE

The thresher sharks are a small family of sharks with three species presently recognized. Threshers are found worldwide in tropical and temperate water, generally in the open ocean environment but occasionally inshore. The threshers are easily recognized by the extremely elongate upper lobe of the caudal (tail) fin. Threshers evidently use this extension for feeding. Often, when threshers are caught on long lines (miles of buoyed, stout line with many baited hooks) the sharks are foul-hooked in this tail fin extension. The reason this occurs is that the thresher shark uses the fin to inflict stunning blows on fish during an attack. There is even a somewhat dubious report of a thresher using its tail fin to kill a seabird, although its usual diet consists of herring, squid, mackerel, and menhaden. Embryonic development in the thresher sharks is interesting because the developing young, after depleting yolk reserves and while still in the uterus, feed on eggs that are produced by the mother.

KEY TO THE SPECIES OF THRESHER SHARKS

1. A. Very large eyes that are oval in shape and directed upward. A distinct groove extending from just behind the eyes along each side of the head, first dorsal fin base closer to pelvics than to pectorals. Bigeye thresher shark, page 94.

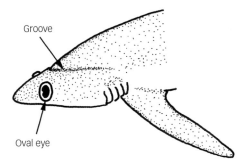

Groove

Oval eye

B. Eyes not enlarged and not directed upward. No groove along each side of head. First dorsal fin base closer to pelvics than to pectorals. Common thresher shark, page 95.

Bigeye thresher shark, *Alopias superciliosus*

CHARACTERS: The bigeye thresher has an elongate upper lobe of the caudal fin similar to the common thresher shark, although in this species the tail can be as long as the rest of the body. The eyes are very large and directed upwards in this species. Additionally, there are two prominent horizontal grooves that intersect just behind the eyes and extend horizontally along the sides to just above the pectoral fins. The first dorsal fin is positioned closer to the tail than to the snout; the second dorsal is minute; the pectoral fins are long and narrow; and the snout is conical and moderately long. The color is gray to purplish gray above and cream below.

BIOLOGY: This is an oceanic species that may occasionally venture inshore. Maximum size is about 460 centimeters (15 feet); males mature at about 270 centimeters (9 feet) and females at about 355 centimeters (11.6 feet). Typically two pups are produced, but litters of four have been reported. Size at birth is from about 100 to 130 centimeters (3.3 to 4.3 feet). This fish feeds on fishes and squids. It is not known to have attacked humans.

DISTRIBUTION AND STATUS: This is an oceanic species that has been ob-served at the surface down to 475 meters (1,560 feet). The bigeye thresher is not a common Gulf of Mexico species but may be captured in the southern and northwestern Gulf. It is found from New York to Florida, including the Gulf of Mexico and south to Brazil. Little is known of its status, but like other large shark species its numbers are likely reduced.

SIMILAR SPECIES: See description for the common thresher.

Common thresher shark, *Alopias vulpinus*

CHARACTERS: The thresher shark's most distinctive characteristic is the extremely oversized upper lobe of the caudal fin, representing about half the total length of the shark. The eyes are moderately large but not directed upward (as in the bigeye thresher). The pectoral fin is long, narrow, and curved. The first dorsal fin is positioned about midway between the tip of the snout and the beginning of the caudal fin, and the second dorsal fin is minute. The color is brown, gray, blue gray, or blackish above and abruptly changes to white below. A white patch extends from the abdomen over the pectoral fin bases.

BIOLOGY: This shark is more likely to be encountered far at sea in the Gulf of Mexico, but young specimens may be found in shallower waters. The maximum reported size of this species is about 760 centimeters (25 feet), although most specimens fall in the 430 to 490 centimeters (14 to 16 feet) range. Males mature at 319 to 420 centimeters (10.5 to 13.7 feet) and females at 376 to 549 centimeters (12.3 to 18.0 feet). Pups are born at about 114 to 150 centimeters (3.7 to 5 feet), and litter size is two pups. This shark feeds on schooling fishes using its long tail as a weapon for stunning prey. Additionally, it feeds on squids, octopi, crustaceans, and seabirds. There are no reliable accounts of it attacking humans, but the large size of this shark makes it a species that should be respected.

DISTRIBUTION AND STATUS: The thresher is found from Newfoundland to Florida, to the Gulf of Mexico, and off of Brazil and Argentina. It is found in shallow surface waters down to 366 meters (1,200 feet). The status of this species is in question because populations of this shark have apparently decreased significantly.

SIMILAR SPECIES: The thresher could be confused with the bigeye thresher, but the bigeye thresher has large upward-directed eyes and distinct grooves on the nape of the neck.

MACKEREL SHARKS
FAMILY LAMNIDAE

The family Lamnidae (the mackerel sharks) contains some of the most feared sharks. Of particular note are the great white and mako sharks. All members of this family are characterized by their distinct tail fins, which have almost equal upper and lower lobes, the torpedo-like shape of the body, the conical snout, the ability to regulate body temperature, flattened keels (support structures) where the tail fin attaches to the body, and large gill openings. All of these characteristics make the mackerel sharks very efficient, high-speed swimmers. It is therefore not surprising that these sharks are near the top of the oceanic food web and are fully capable of preying upon swordfish, seals, dolphins, and other sharks. Mackerel sharks are found worldwide in temperate and tropical waters, but the largest numbers of these sharks are found in cold water areas.

KEY TO THE SPECIES OF MACKEREL SHARKS

1. **A.** Teeth in upper jaw triangular in shape with heavy serrations, origin of first dorsal just anterior of rear corner of pectoral fin. White shark, page 98.

   B. Teeth in upper jaw slender, bladelike, and not serrated. Origin of first dorsal behind rear corner of pectoral fin. Go to #2.

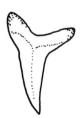

2. **A.** Underside of snout and area around mouth pigmented. Pectoral fins long, more than 18 percent of total length of shark. Longfin mako, page 101.

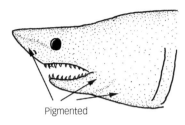

Pigmented

B. Underside of snout and area around mouth white. Pectoral fins short, about 17 percent of total length of shark. Shortfin mako, page 99.

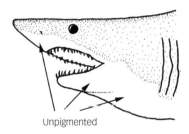

Unpigmented

Great white shark, *Carcharodon carcharias*

CHARACTERS: This is a very large shark that has a conical snout, a bullet-shaped body, and a lunate (shaped like a crescent moon) caudal fin. The eyes are black and the teeth are large, triangular, and heavily serrated. The color ranges from gray to brown or black above, abruptly changing to white below. There is a black spot near the rear of the pectoral fin base.

BIOLOGY: The great white shark is evidently a rare visitor to the waters of the Gulf of Mexico. It is known, however, that this shark typically makes its appearance during the winter months (January to April). Most of the white sharks observed in the Gulf of Mexico were seen off of the west coast of Florida from Clearwater to Cape Sable in water temperatures of 18.7 to 21.6°C (65.7 to 70.9°F). I have personally examined several white sharks landed at fish houses in and around John's Pass, Florida. Sharks captured in the Gulf ranged from 185 to 472 centimeters (6.1 to 15.5 feet) total length. Male white sharks mature at about 275 centimeters (13 feet), and females mature at about 400 centimeters (9 feet). The largest recorded white shark was 720 centimeters (23.6 feet). Size at birth is about 120 centimeters (4 feet), and litters of up to fourteen pups have been recorded. White sharks in the Gulf prey upon other sharks, bottlenose dolphins, and fish, but will likely take a variety of prey items. This shark is known to attack and consume humans.

DISTRIBUTION AND STATUS: In the western Atlantic it occurs from Newfoundland to Argentina, including the Gulf of Mexico. In the Gulf it has been captured at 20 to 164 meters (65 to 538 feet) depth. The white shark is protected in some waters and is likely much reduced in numbers wherever it occurs.

SIMILAR SPECIES: White sharks can be confused with shortfin or longfin makos. However, the teeth in both the longfin and shortfin mako are blade-like, slender, and unserrated. The white shark tooth is triangular and heavily serrated.

Shortfin mako shark, *Isurus oxyrhynchus*

CHARACTERS: The shortfin mako is one of the most striking sharks found in the Gulf of Mexico. It is deep blue above and white below, and the body is slender and very much torpedo shaped. The snout is long, acutely pointed, and conical. The teeth are long, slender, very bladelike, and unserrated and point backwards. The eyes are large and black. The first dorsal fin is relatively large, and the second dorsal and anal are very small. The underside of the snout around the mouth is white. The tail is lunate in shape. The pectoral fin length is less than the head length.

BIOLOGY: The shortfin mako is an oceanic species rarely appearing inshore. It apparently prefers the sunlit surface waters of the oceans but has been captured inshore. It is perhaps the fastest swimming of all sharks and, when hooked by fishers, will make spectacular leaps from the water. The maximum size is about 400 centimeters (13 feet); males mature at about 200 centimeters (6.5 feet) and females at about 280 centimeters (9.2 feet). Size at birth is about 60 to 70 centimeters (2 to 2.3 feet), with as many as eighteen pups in a litter. This species feeds on other sharks, fishes, squids, and dolphins. It has attacked humans and should be considered dangerous.

DISTRIBUTION AND STATUS: In the western Atlantic the shortfin mako is found from the Gulf of Maine to southern Brazil and includes the Gulf of Mexico. I have personally collected shortfin mako sharks from the northern Gulf just south of Mobile Bay, Alabama. The shortfin mako shark is likely much reduced in numbers.

SIMILAR SPECIES: The shortfin mako shark can be distinguished from the white shark by its more slender, unserrated teeth. The shortfin is easily distinguished from the longfin mako by the dusky coloration around the mouth of the longfin (the shortfin has no such coloration).

THE SPEEDING SHORTFIN MAKO

The shortfin mako has to be my favorite shark. This fish is shaped like a torpedo and is a beautiful cobalt blue above, almost exactly matching the ocean's blue color, and snow white below. Shortfin makos are the race horses of the oceans. Everything about them—torpedo-like body shape, powerful tail, ability to retain heat, and muscular body—belies their swimming ability. Makos love to leap from the water when hooked, and based on the height above the water that mako sharks rise, their speed has been estimated to be about twenty-two miles per hour. This may not seem very fast but consider that fish swim through a medium, water, that is about sixty times more viscous (thick) and eight hundred times more dense than air. A 22 mph speed in water is equivalent to a 100 mph speed in air. I have only caught a few makos in the Gulf of Mexico, but one mako capture was very memorable. While conducting shark research, we brought a large, very lively mako shark alongside the research vessel. For safety reasons, we always dispatched large sharks with a shotgun before bringing them on board. I held the line while a fellow researcher took aim. The boat rolled just as he pulled the trigger, completely missing the shark and striking the side of the, thankfully, steel-hulled vessel. The mako was understandably surprised and burst away across the surface of the water like a torpedo, as fast as a mako can go. I was still holding the line and obtained a severe rope burn before I could release it. Before bringing the shark back in we quickly went below deck to find that the shot had not penetrated the hull. I hope to never have another speeding mako shark on the end of a hand line.

Longfin mako shark, *Isurus paucus*

CHARACTERS: The longfin mako has pectoral fins longer than the head length; a long, pointed snout; a torpedo-shaped, slender body; a lunate tail; a large first dorsal and very small second dorsal; and teeth that are slender, unserrated, and rather bladelike. There are prominent keels (flattened, shelf-like extensions) on either side of the body where the caudal fin attaches. The longfin mako is intense blue on the back and sides, and the belly is bluish gray to dusky. The underside of the snout and the area around the mouth is dusky in color

BIOLOGY: The longfin mako is not common in the Gulf of Mexico but may be encountered, particularly in the southern Gulf. The first record of a longfin mako in the Gulf of Mexico was in 1985 from about eighty miles south of Panama City, Florida. The species has been captured from the surface to about 220 meters (720 feet) depth. It feeds on fishes and squids. The maximum size is about 417 centimeters (13.7 feet); females and males mature at about 245 centimeters (8 feet). Size at birth is about 97 centimeters (3.2 feet), and litter size is two pups. Although no attacks on humans have been reported, this shark is large and very active and should be regarded with caution.

DISTRIBUTION AND STATUS: The species ranges in the western Atlantic from Florida to Cuba, including the Gulf of Mexico, to southern Brazil. Although poorly known, the numbers of longfin makos have probably been declining in the Gulf of Mexico along with all other larger species of sharks.

SIMILAR SPECIES: See description of the shortfin mako.

DOGFISH SHARKS
FAMILY SQUALIDAE

Although there are only a few species of dogfish in the shallow waters of the Gulf of Mexico, this family contains a great variety of predominantly small sharks. Dogfish are best identified by their two similar-sized dorsal fins that may or may not have spines preceding them and no anal fin. Some of the deep-water species have light-emitting organs along the sides and belly. The cookie cutter sharks, *Isistius brasiliensis* and *Isistius plutodus*, are parasitic on larger fish. Dogfish are often found in large schools, and some species constitute an important commercial fishery. Dogfish dine upon a great diversity of animals, including fish, squid, octopus, and shrimp. The size of the different kinds of adult dogfish ranges from seven inches to twenty feet. Although there are some twenty species in the Gulf of Mexico, only a single species, the Cuban dogfish, is likely to be encountered by the average naturalist or fisher. If the reader requires identification of any of the other nineteen species of dogfish found in the deep waters of the Gulf, I suggest obtaining *Fishes of the Gulf of Mexico* (volume 1) by John D. McEachran and Janice D. Fechhelm.

Cuban dogfish shark, *Squalus cubensis*

CHARACTERS: The Cuban dogfish is a small, slender, gray shark with black-tipped dorsal fins. The rear edges of the pectorals, pelvics, and caudal fins are white. Each dorsal is preceded by a stout spine, the first dorsal being larger than the second. The teeth are alike in each jaw, being small with a deeply notched single cusp that is very oblique. The eyes are relatively large and the iris is green. The shark is dark gray with paler gray below.

BIOLOGY: The Cuban dogfish is a bottom-dwelling shark found in large schools, from about 60 to 380 meters (200 to 1,250 feet) depth. It feeds on bottom organisms such as fish, crabs, and shrimps. There is a large parasitic isopod (crablike organism) that is found in the mouth of this species. The maximum size is about 110 centimeters (3.6 feet); both males and females mature at about 50 to 75 centimeters (1.6 to 2.5 feet). Litter size is about ten pups. This is a harmless species.

DISTRIBUTION AND STATUS: In the western Atlantic it is found from North Carolina to Florida, including the Gulf of Mexico, Cuba, and the Antilles, and south to Brazil. The populations of this species are likely secure, although little information is available.

SIMILAR SPECIES: Although there are several species of deep-water sharks in this family, the Cuban dogfish is the only species of this family that could possibly be encountered in relatively shallow water, and it should not be easily confused with other species.

SAND TIGER SHARKS
FAMILY ODONTASPIDIDAE

The sand tiger sharks are a relatively small group of sharks with eight species presently recognized but only two species in the Gulf of Mexico. All of the species of sand tiger sharks closely resemble one another. Their characteristics include the first dorsal fin set far back on the body; the first and second dorsal fins similar in size; long, slender, unserrated teeth; and a generally "stocky" appearance to the body. Probably the most impressive feature of sand tiger sharks is the extremely "toothy" grin that these sharks possess. The teeth almost appear to be growing from the mouth in an uncontrolled fashion. All sand tiger sharks in the western Atlantic are protected by the National Marine Fisheries Service, and their commercial or recreational harvest is prohibited.

KEY TO THE SPECIES OF SAND TIGER SHARKS

1. **A.** Origin of the dorsal fin (a.) is over the pectoral fin; first dorsal fin is larger than the second dorsal; the snout is conical (round as opposed to flattened). Bigeye sand tiger. Go to page 106.
   B. Origin of the dorsal fin (a.) is about midway between pectoral and pelvic fins; the first and second dorsals are similar in size; the snout is flattened. Sand tiger shark. Go to page 105.

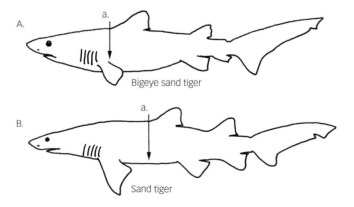

A.

a.

Bigeye sand tiger

B.

a.

Sand tiger

Sand tiger shark, *Eugomphodus taurus*

CHARACTERS: The sand tiger has a "stocky" appearance; a short pointed snout; protruding, bladelike teeth; small, similarly sized dorsal and anal fins; and a caudal fin that is large with a small ventral lobe. The first dorsal fin originates about midway between the pectoral and pelvic fins. The color is pale brown or gray, lighter below, with yellowish brown to reddish spots on the upper surfaces and sides.

BIOLOGY: The sand tiger is most easily recognized by its extremely snaggled teeth that project from the mouth. Their seemingly fierce appearance make them popular for display at public aquaria. The species prefers shallow, warm water near the bottom, but may also be seen in mid-water. This shark feeds on fishes, including other sharks and rays, squids, crabs, and lobsters. Developing embryos feed on ova produced by the mother and cannibalize their siblings such that only one pup per uterus is produced (litter size = two). Gestation is about eight to nine months, and size at birth is about 100 centimeters (3.3 feet). Maximum size is about 320 centimeters (10.5 feet); females mature at about 220 to 300 centimeters (7.2 to 9.8 feet) and males at about 220 to 257 centimeters (7.2 to 8.4 feet). It has been reported that this species has attacked divers, but in many situations it seems to completely ignore humans. However, large individuals should be treated with caution.

DISTRIBUTION AND STATUS: This species is found in the western Atlantic from the Gulf of Maine to southern Brazil including the Gulf of Mexico. It may be found from very shallow water near shore to 191 meters (630 feet) depth. The numbers of sand tiger sharks have declined precipitously in recent years and are likely in trouble in western Atlantic waters. This species is protected in the Atlantic by the National Marine Fisheries Service.

SIMILAR SPECIES: The sand tiger could be confused with the less common bigeye sand tiger. However, the bigeye sand tiger has a larger eye, and the second dorsal originates well behind the origin of the pelvic fins. The sand tiger has a second dorsal well ahead of the pelvic fins.

Bigeye sand tiger shark, *Odontaspis noronhai*

CHARACTERS: The sand tiger is most easily recognized by their extremely snaggled teeth that project from the mouth. This shark has large oval eyes, a bulbous snout, and a large, low caudal fin. The teeth have a slender, bladelike cusp with a lateral cusp on each side. The first dorsal fin is larger than the second, and the anal fin is small. The color is uniformly chocolate brown, and all fins except the pectorals have a dark edge along their rear margins.

BIOLOGY: This is a poorly known inhabitant of deep water between 600 and 1,000 meters (1,970 to 3,300 feet) depth. Little is known of its biology. Maximum size is about 360 centimeters (12 feet); females mature at about 326 centimeters (11 feet), and males mature at about 326 to 342 centimeters (11 to 11.2 centimeters). Reproduction is likely similar to that in the sand tiger shark wherein the embryos feed upon eggs produced by the mother and cannibalize their siblings. Litter size is likely two.

DISTRIBUTION AND STATUS: In the Gulf of Mexico this species has been recorded from a single specimen captured off Texas. Too little information is available to determine its status.

SIMILAR SPECIES: See sand tiger description.

WHALE SHARKS
FAMILY RHINCODONTIDAE

The whale shark is the largest and perhaps the most impressive fish in the world. Growing to sixty-five feet in length, it is perhaps a blessing that these sharks are harmless filter feeders; that is, they feed by drawing large volumes of water into the mouth and filtering out the small marine creatures that exist there. If these sharks fed on humans, the oceans would definitely be a less-hospitable place. There is only a single representative of this family. In the United States the whale shark is protected by the National Marine Fisheries Service, and their commercial or recreational harvest is prohibited. The whale shark is also CITES listed, meaning that international trade in this species is prohibited.

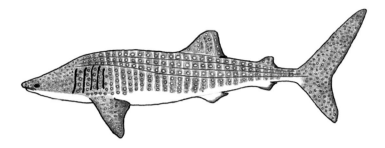

Whale shark, *Rhincodon typus*

CHARACTERS: The whale shark has an upper surface that may be grayish, bluish, or brownish with cream-colored white spots between pale, vertical, and horizontal stripes. The body is streamlined, and the head is broad and flattened. The caudal is lunate (shaped like a crescent moon). The mouth is located at the tip of the snout, and the gills slits are very large.

BIOLOGY: Whale sharks are typically found far out at sea in warm temperate and tropical waters, cruising close to the ocean's surface. Whale sharks are not uncommonly sighted in the Gulf of Mexico, often accompanied by large schools of tuna. Whale sharks apparently return to the same areas year after year, and their regular appearance has become a tourist attraction in some areas. These sharks filter feed small organisms from the ocean's waters. Litter size is over three hundred pups born at about 55 to 64 centimeters (1.8 to 2.1 feet). Maximum size is about 18 meters (59 feet). There is very little information available concerning their biology. This species has been reported to have rammed boats but is not dangerous. However, any very large animal should be treated with caution.

DISTRIBUTION AND STATUS: In the western Atlantic they are found from New York to central Brazil. Although most commonly observed far at sea in the Gulf of Mexico, in some areas of the world this species comes close to shore. Specimens have been observed around the gas rigs off the Mississippi coast, south of the mouth of the Mississippi River, and around the Flower Garden Reefs off eastern Texas. The numbers of whale sharks have declined, and they are now protected by U.S. and international law.

SIMILAR SPECIES: This shark could not be confused with any other species.

SMOOTHOUND SHARKS
FAMILY TRIAKIDAE

The smoothounds are a fairly large family of sharks that are closely related to the family Carcharhinidae (the requiem sharks). There are about thirty-nine species in this family, although only two species are found in the Gulf of Mexico. They are identified by their slender body, small flattened teeth set close together, second dorsal fin about three-fourths the size of the first dorsal, and large narrow eyes. The smoothounds are found in shallow tropical and temperate waters and are bottom oriented. Some of the species of this family survive well in captivity and can be seen in public aquaria.

KEY TO THE SPECIES OF SMOOTHOUND SHARKS

1. **A.** Upper labial furrows (creases around corners of mouth) (a.) longer than lower labial furrows (b.) Lower lobe of caudal fin broadly rounded. Smooth dogfish. Go to page 110.
   B. Upper labial furrows (a.) shorter than lower labial furrows (b.) Lower lobe of caudal fin sharp, pointed, and directed rearward. Florida smoothound. Go to page 111.

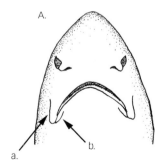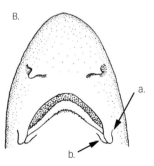

Smooth dogfish shark, *Mustelus canis*

CHARACTERS: The smooth dogfish is a small species with similarly sized first and second dorsal fins, a moderately developed lower caudal fin lobe, large, elongate eyes, and a prominent spiracle. The trunk is slender and tapers rearward. The teeth are bluntly rounded. The color is olive gray to brown above and pale gray to off-white below. The margins of the fins may be paler in color.

BIOLOGY: The smooth dogfish is typically found near the bottom, and large numbers of them may be encountered in some areas of their range. Smooth dogfish feed on various fish species, lobsters, crabs, squids, and mollusks. Maximum size is about 150 centimeters (5 feet); females mature at about 90 centimeters (3 feet) and males at about 82 centimeters (2.7 feet). Litter size ranges from four to twenty pups, born at about 34 to 39 centimeters (1.1 to 1.3 feet). This species is harmless.

DISTRIBUTION AND STATUS: Found in the western Atlantic from the Bay of Fundy to southern Brazil, including the Gulf of Mexico and the Greater and Lesser Antilles. Its depth distribution is from near shore to 200 meters (660 feet). The populations of this species are probably secure.

SIMILAR SPECIES: See description of the Florida smoothhound.

Florida smoothound shark, *Mustelus norrisi*

CHARACTERS: The Florida smoothound is a small species with similarly sized first and second dorsal fins, a moderately developed lower caudal fin lobe, large, elongate eyes, and a prominent spiracle. The trunk is slender and tapers rearward. The teeth are bluntly rounded. The color is gray above and pale gray to off-white below. The margins of the fins may be paler in color.

BIOLOGY: This species prefers muddy or sandy bottoms and is often taken in large numbers as by-catch of commercial fisheries in the northern Gulf. It feeds on other fishes, crabs, and shrimps. Litter size ranges from seven to fourteen pups, born at about 30 centimeters (1 foot). Maximum size is about 110 centimeters (3.6 feet); females mature at about 65 centimeters (2.1 feet) and males at about 58 centimeters (1.9 feet). This is a harmless species.

DISTRIBUTION AND STATUS: In the western Atlantic this species is found from Florida to southern Brazil, including the Gulf of Mexico. It is relatively common in the northern Gulf of Mexico to about 80 meters (260 feet) depth. The populations of this species are probably secure.

SIMILAR SPECIES: Distinguishing this species from the smooth dogfish can be difficult. However, the Florida smoothound has nostrils that are closer together (distance between nostrils 1.2 to 2.4 percent of shark length) than the smooth dogfish (distance between nostrils 2.7 to 3.6 percent of shark length).

NURSE SHARKS
FAMILY GINGLYMOSTOMATIDAE

While there are many species in this family worldwide, there is only a single representative in the western Atlantic. The nurse sharks are sluggish, bottom-dwelling fish, preferring shallow, warm-temperate to tropical waters. There are two dorsals; the head tends to be large; the caudal has a poorly developed lower lobe; and the teeth are small. As far as sharks go, this family contains some of the most colorful species of sharks. Some species retain the eggs internally until they hatch whereas others are egg layers.

Nurse shark, *Ginglymostoma cirratum*

CHARACTERS: The nurse shark has a long barbel on each nostril; the mouth is far forward on the head; the eyes are small; the head is blunt; all the fins are very rounded; the second dorsal is one-half to two-thirds as large as the first; the teeth are very small; the caudal has a poorly developed lower lobe. The color is yellow to yellow-green or reddish brown in adults. The young have dark spots and saddles on a lighter background.

BIOLOGY: The nurse shark is common in shallow water and, being nocturnal, can be observed resting motionless on the bottom during the day. I have captured many small nurse sharks while snorkeling in the Florida Keys by pulling them from under coral reef ledges. They are common around reefs, in mangrove areas, and on sand flats. They can sometimes be observed mating in shallow water. Maximum size is about 430 centimeters (14 feet); females mature at about 235 centimeters (7.7 feet) and males at about 225 centimeters (7.4 feet). Litters of twenty-one to twenty-eight pups have been reported, and size at birth is about 27 centimeters (0.9 feet). Nurse sharks feed on a variety of bottom organisms including lobsters, shrimps, crabs, squids, octopi, snails, and fishes. They may bite if harassed, and a large individual could cause serious injury.

DISTRIBUTION AND STATUS: The nurse shark is found in the western Atlantic from Rhode Island to southern Brazil and is common off of Florida. They are normally found in shallow water to about 12 meters (40 feet) depth. The populations of this species are probably secure, although little information exists.

SIMILAR SPECIES: There are no similar species in the Gulf of Mexico.

BASKING SHARK
FAMILY CETORHINIDAE

This family contains a single representative, the basking shark.

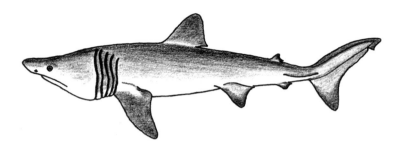

Basking shark, *Cetorhinus maximus*

CHARACTERS: Basking sharks are enormous, with extremely long gill slits, a long, conical snout, and a lunate caudal fin. The first dorsal is large and the second much smaller. The color is grayish brown, blue-gray to blackish above and slightly lighter below. There may be white bands or patches on the snout and belly.

BIOLOGY: The basking shark is a warm-temperate species typically observed near the surface, filter feeding small organisms from the sea by slowly swimming with its mouth open. This is a poorly known species. It grows to a maximum size of about 1,000 centimeters (33 feet); females mature at about 810 to 980 centimeters (27 to 32 feet) and males mature at about 400 to 500 centimeters (13 to 16.4 feet). Litter size is unknown but pups are born at about 165 centimeters (5.4 feet). This shark is harmless to humans.

DISTRIBUTION AND STATUS: This is a very uncommon shark in the Gulf of Mexico and has been observed off of the east coast of Florida and in the northern Gulf of Mexico. In the western Atlantic it ranges from Newfoundland to Florida and from southern Brazil to southern Argentina. Little information is available regarding status.

SIMILAR SPECIES: There are no similar species in the Gulf of Mexico.

CATSHARKS
FAMILY SCYLIORHINIDAE

The catsharks are a group of sharks that are most commonly observed in very deep water in the Gulf of Mexico. The catsharks are small with maximum sizes being less than about 50 centimeters (20 inches). Although sharks in general tend to be fairly drab, these are some of the most colorful sharks in our area. They are distinguishable from other sharks in that they are elongate, have fairly large, slitlike eyes, and the caudal fin is low and not well developed. I have trawled many catsharks, and they are not uncommonly taken on the royal red shrimp grounds in the Gulf. These sharks prefer the bottom and lay eggs in a tough egg case similar to those found in the skates. Worldwide there are about one hundred species of catsharks, but there are only six in the Gulf, and of those, only two species have some chance of being taken by the average fisher or naturalist.

KEY TO THE SPECIES OF CATSHARKS

1. **A.** Dark brown to light brown weblike or chainlike color pattern on a light brown to tan to yellowish background, the top of the caudal fin (a.) does not have enlarged denticles (scales). Chain dogfish. Go to page 117.

B. Distinct to diffuse, grayish to brownish blotches or saddles, some outlined with white, the top of the caudal fin (b.) has enlarged denticles (scales). Marbled catshark. Go to page 116.

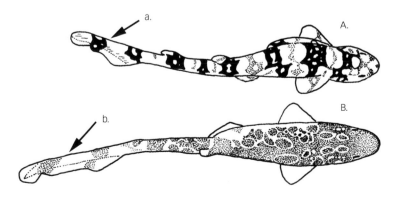

Marbled catshark, *Galeus arae*

CHARACTERS: The marbled catshark is brownish to grayish with dark spots and saddles, many of which are outlined in white. The background color is pale yellowish brown. There is a fine pattern of spots and blotches, particularly on the head. This shark has two similar-sized, small dorsal fins, the origin of the first dorsal is over the middle of the base of the pelvic fins. The teeth have a slender central cusp, with one, two, or three lateral cusps on each side. The caudal is small and has a series of enlarged denticles (scales) along the upper edge.

BIOLOGY: The marbled catshark is a deep-water shark that prefers the bottom. It likely feeds on a number of bottom-dwelling organisms but is known to feed on deep-water shrimps. Maximum size is 43 centimeters (1.4 feet); males mature at 27 to 36 centimeters (10 to 14 inches) and females at 26 to 43 centimeters (10 to 17 inches). Eggs are laid in tough egg cases. This shark is harmless to humans.

DISTRIBUTION AND STATUS: This shark is often trawled from the deep-water shrimp grounds in the Gulf of Mexico. It is a deep-water species, found from 292 to 732 meters (950 to 2,400 feet) depth. In the western North Atlantic it is found from South Carolina to Florida, including the northeastern Gulf and from Belize to Costa Rica. The status of this species in the Gulf of Mexico is unknown.

SIMILAR SPECIES: This species could be confused with the chain dogfish, but the color patterns and the positions of the first dorsal relative to the pelvics easily separate the two.

Chain dogfish shark, *Scyliorhinus retifer*

CHARACTERS: The chain dogfish's most notable characteristic is its color. It has a bold color pattern that can be described as chain or weblike in appearance. The pattern is black to dark brown on a lighter brown, tan, or yellowish background. The chain dogfish has two similarly sized dorsal fins placed far back on the tail; the origin of the first dorsal is over the free rear edge of the pelvics; the snout is short; and the caudal is poorly developed. The teeth, similar in both jaws, have a triangular central cusp with a smaller lateral cusp on each side (sometimes two lateral cusps). There are generally three or four functional rows of teeth.

BIOLOGY: This small shark prefers the bottom where it feeds upon shrimps, crabs, and small fishes. The chain dogfish lays eggs in a tough egg case, and the newly hatched pups are about 10 centimeters (4 inches) long. The number of eggs laid in a season is not known. Maturity is reached at about 37 to 41 centimeters (1 to 1.3 feet) in males and 35 to 47 centimeters (1.1 to 1.5 feet) in females. Maximum size is about 47 centimeters (1.5 feet). This shark is harmless to humans.

DISTRIBUTION AND STATUS: This shark is not uncommonly trawled from the royal red shrimp grounds in the Gulf of Mexico. It is likely found over most of the Gulf in waters of about 75 to 550 meters (250 to 1,800 feet) depth. In the western Atlantic it ranges from southern New England to Florida and to the northern Gulf of Mexico and south to the western Caribbean, the Yucatan, and Nicaragua. There is little information available concerning the populations of this shark.

SIMILAR SPECIES: In the Gulf of Mexico this species could be confused only with the marbled catshark, but their color patterns differ significantly, making them easy to distinguish.

GOBLIN SHARK
FAMILY MITSUKURINIDAE

This family contains a single species, the goblin shark, Mitsukurina owstoni.

Mouth open

Goblin shark, *Mitsukurina owstoni*

CHARACTERS: The goblin shark is characterized by an elongate, flat, blade-like snout, a very small eye, a long pelvic fin base, and pincer-like, highly pro-trusible, and expansible jaws. There are two dorsal fins. The teeth are similar in both jaws, with slender, unserrated, thornlike cusps on broad bases with smaller lateral cusps on some teeth. The color is pinkish gray.

BIOLOGY: The goblin shark is arguably the rarest (and perhaps ugliest) shark in existence, with only a single specimen captured in the Gulf of Mexico. The shark was captured about 190 km (120 miles) south of Pascagoula, Mississippi, at about 1,000 meters (3,300 feet) depth on the bottom. The shark was estimated to be 540 to 617 centimeters (17.7 to 20.2 feet) in length and is believed to be the largest specimen ever captured. This species is a bottom dweller and is known to feed on squid and likely a variety of other prey items. Almost nothing is known of its biology. This shark is too rare to ever be a threat to humans.

DISTRIBUTION AND STATUS: The goblin shark is probably found worldwide in deep ocean waters, but at present it has only been captured off the coasts of Japan, South Africa, Australia, French Guiana, Europe, and New Zealand and in the Indian Ocean and the Gulf of Mexico. Depth of capture was from 40 to 1,300 meters. There is no information on the status of goblin shark populations.

SIMILAR SPECIES: The goblin shark could not be confused with any other species.

Skate and Ray Identification

KEY TO THE FAMILIES OF SKATES AND RAYS OF THE GULF OF MEXICO

(Choose the diagram that best describes your fish.)

1. BODY SHAPE

A. Body sharklike, toothed saw extending from rostrum. Sawfishes, page 123.

B. Elongate body. Guitarfish, page 126.
C. Body largely flattened and/or with well-defined "wings." Go to #2.

2. DISC SHAPE

A. Disc is round or angular but not with birdlike wings. Go to #3.

B. Disc has pointed, winglike fins (see diagrams below)

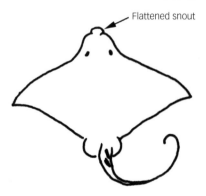

Snout with hornlike fins. Manta rays, page 159.

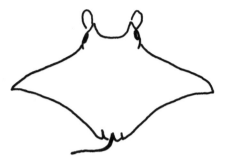

Ducklike, flattened snout. Eagle rays, page 154.

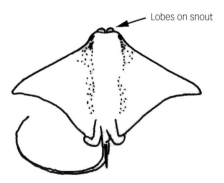

Snout with 2 lobes. Cownosed ray, page 157.

Very wide, expanded wings. Butterfly rays, page 142.

3. STINGING SPINE(S) PRESENT/ABSENT

A. There is (are) stinging spine(s) present on the upper surface of the tail.
Disc is very round; color is dark reticulations on a light background. Round stingray, page 131.

Spine

Disc is more angular but definitely not with "wings" and not round. Tail is very long. Stingrays, page 133.

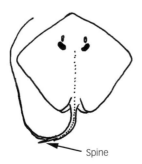

Spine

B. There is no stinging spine present on the tail. Go to next page.

The tail is developed and possesses thorns. There are two dorsal fins. The skates, page 144.

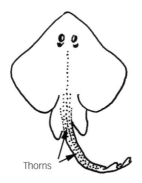

Thorns

Electric organs

The tail has no thorns. The disc has two kidney-shaped electric organs barely visible through the skin. Electric rays, page 128.

SAWFISHES
FAMILY PRISTIDAE

Because of their sharklike appearance, people often refer to the sawfishes as sharks. However, they are actually related to the skates and rays because the gills are ventral, the eyes are on top of the head, and the pectoral fins are attached to the head. These interesting animals are easy to recognize thanks to the toothed rostral process (the saw) that extends from the head. The teeth in the saw are firmly attached along each side of the structure. The saw is used for feeding by slashing through schools of fish and digging in the bottom to locate prey. The sawfishes have two well-developed dorsal fins, and the head is flattened. The sawfishes are typically found in tropical shallow marine areas but are known for their ability to live in completely freshwater. At one time these fish were apparently very common in the Gulf of Mexico, particularly in the Florida Keys, but have since become rare. Sawfish numbers have declined precipitously in the United States, and the smalltooth sawfish is now protected under the Endangered Species Act. The fact that the toothed saw is easily entangled in fishing gear has probably contributed to this animal's decline. Two species are found in the Gulf of Mexico.

KEY TO THE SPECIES OF SAWFISHES

1. A. Rostrum (saw) with 16 to 22 teeth along one side. Largetooth sawfish, page 125.

B. Rostrum (saw) with 23 to 34 teeth along one side. Smalltooth sawfish, page 124.

Smalltooth sawfish, *Pristis pectinata*

CHARACTERS: The smalltooth sawfish has a greater number of teeth in the saw (23 to 34) than does the largetooth sawfish, and its caudal fin has no lower lobe. Its color on the upper surface is dark gray to blackish brown, and this fades to whitish or yellowish below. Like the largetooth sawfish, this species has two dorsal fins of similar size, and its head is flattened.

BIOLOGY: The smalltooth sawfish has historically been the more common of the two species found in the Gulf of Mexico. The smalltooth grows to a maximum length of about 550 centimeters (18 feet); it matures at about 300 centimeters (10 feet); and its size at birth is about 60 centimeters (2 feet). Development is by aplacental viviparity. Twenty young have been observed in a litter, and young are born in spring and summer in south Florida. Young sawfish in southwest Florida are typically found in extremely shallow water. This species feeds on bottom-dwelling organisms such as shrimps and crabs and on schooling fishes such as mullet, herring, and menhaden. The saw could inflict serious injury, but otherwise it is harmless to humans.

DISTRIBUTION AND STATUS: The smalltooth is a shallow-water species, and in the western Atlantic it ranges from New Jersey, throughout Florida and the Gulf of Mexico, south to Brazil. The populations in the Gulf are in serious decline, and the smalltooth sawfish is now protected under the Endangered Species Act. Most specimens reported now come from southwest Florida.

SIMILAR SPECIES: See description for largetooth sawfish.

Largetooth sawfish, *Pristis pristis*

CHARACTERS: The largetooth sawfish is thus named because it has fewer and larger teeth on the saw than the smalltooth sawfish. This species has sixteen to twenty-two teeth along one side of the saw, and its caudal (tail) fin has a lower lobe. There are two well-developed dorsal fins, and the head is flattened. Freshly captured animals may be dark gray, golden brown, or pale yellow on the upper surface and whitish underneath.

BIOLOGY: The largetooth sawfish grows to a maximum length of about 650 centimeters (21 feet) total length. A specimen from Texas was reported to be about 570 centimeters (17.5 feet) and weighed 1,200 pounds. This species has live birth. Gestation is about five months, and the young are born at about 60 centimeters (2 feet) total length. It is found in shallow coastal waters and can remain indefinitely in freshwaters. As a matter of fact, it has been taken 450 miles up the Amazon River. This fish is apparently restricted to waters that do not fall below about 20°C (68°F). The largetooth sawfish poses little threat to humans, although its powerful slashing saw could inflict inadvertent serious injury. Prey items include benthic invertebrates (crabs, shrimps, etc.) and fishes.

DISTRIBUTION AND STATUS: In the western Atlantic, the largetooth sawfish is distributed throughout the Gulf of Mexico from southern Florida to Texas and the Caribbean Sea. The species is apparently found exclusively in shallow waters. The populations of largetooth sawfish have been greatly reduced, and this species may be in trouble.

SIMILAR SPECIES: The closely related smalltooth sawfish (see below) can be easily separated from this species by the greater numbers of teeth in the saw of the smalltooth (25 to 32) and the fact that the smalltooth has no lower lobe on the caudal. The saw shark, a deep-water species, is similar, but the teeth in the saw are more loosely attached and the saw shark has two barbels on the underside of the saw.

GUITARFISH
FAMILY RHINOBATIDAE

Only a single member of this family of about forty species is found in the Gulf of Mexico. The guitarfish is an elongate ray possessing a wedge-shaped snout with a rounded tip. The head is flattened while the tail is robust, rounded above, and flattened below. The two dorsal fins are similar in size and shape. The guitarfish is bottom oriented and is frequently observed lying quietly on the bottom or half-buried in the sand. There are blunt thorns present over much of the body.

Atlantic guitarfish, *Rhinobatos lentiginosus*

CHARACTERS: The color of the Atlantic guitarfish is gray to olive brown or chocolate brown above with much of the dorsal surface covered with many small, whitish dots. However, specimens from the northwestern Gulf may have few or no spots. The body is covered with blunt thorns, and a line of larger thorns extends down the centerline of the animal from just behind the eyes to the first dorsal fin. There are a few enlarged thorns on the tip of the snout. The disc has been described as "heart" shaped.

BIOLOGY: The guitarfish grows to a maximum size of about 75 centimeters (2.5 feet). Males mature at about 50 centimeters (1.7 feet). Females likely mature at a slightly larger size. A female containing six young has been reported, and litter size is thought to be fewer than ten. The size at birth is about 20 centimeters (8 inches). Development is via aplacental viviparity. The species preys upon bottom-dwelling organisms (mollusks, crustaceans) and small fishes. It is harmless to humans.

DISTRIBUTION AND STATUS: This species occurs throughout the Gulf of Mexico. It is a shallow-water inhabitant, occurring out to about 100 meters (about 330 feet) depth. There is too little information available to draw conclusions regarding its status.

SIMILAR SPECIES: The long, wedge-shaped snout, the muscular tail, and elongate body distinguish this fish from all other ray species in the Gulf of Mexico.

ELECTRIC RAYS
FAMILIES TORPEDINIDAE
AND NARCINIDAE

Two families of electric rays are found in the coastal waters of the Gulf of Mexico, with a single species in each family. The electric rays are thus named for their amazing ability to generate a strong electric discharge that is used to stun prey and for defense. The large, paired, kidney-bean-shaped electric organs of these fish are visible through the skin on the dorsal surface of the animal. The disc of these rays is very thick at the margins and the tail is short and thick with two dorsal fins.

KEY TO THE SPECIES OF ELECTRIC RAYS

1. **A.** The margin across the front of the disc is rounded. Family Narcinidae, lesser electric ray, page 129.

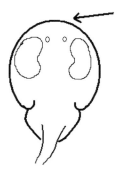

   B. The margin across the front of the disc is relatively straight. Family Torpedinidae, torpedo ray, page 130.

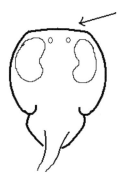

Family Narcinidae

Lesser electric ray, *Narcine brasiliensis*

CHARACTERS: The lesser electric ray varies from a uniform dark brown to light brown or reddish orange with irregular dark rings, ovals, or loops. The disc in this species is rounded, and the snout is long. The first and second dorsal fins are similar in size.

BIOLOGY: The lesser electric ray is the common electric ray found along the beaches and coastal waters of the Gulf of Mexico, and swimmers may tread upon them for they prefer to bury themselves in the sand. Its interesting color pattern makes it one of the most handsome rays in the Gulf. The size at birth of this species is about 12 centimeters (5 inches); males mature at 22.5–25 centimeters (9–10 inches); females mature at about 27 centimeters (10 inches). Maximum size is around 45 centimeters (18 inches). Development is via aplacental viviparity, and maximum litter size is fifteen. The majority of embryos in a litter tend to be of one sex. It feeds on worms, eels, anemones, and crustaceans. It is a nocturnal species. Its electric organ produces 14 to 37 volts. The lesser electric ray is harmless.

DISTRIBUTION AND STATUS: In the western Atlantic, this species is distributed from North Carolina, throughout the entire Gulf of Mexico and the Caribbean, to Brazil. It is a shallow-water species. Little information is available concerning population size, but it is not believed to be in peril at this time.

SIMILAR SPECIES: See description of torpedo ray.

Family Torpedinidae

Atlantic torpedo ray, *Torpedo nobiliana*

CHARACTERS: The torpedo ray is dark brown to purplish brown or nearly black with occasional darker spots. The disc in the Atlantic torpedo ray is subcircular; that is, it is wider than it is long. There is a very short snout, and the margin of the disc in front of the eyes is nearly straight. The caudal fin is very large.

BIOLOGY: The torpedo ray is not commonly encountered in the Gulf of Mexico. Small fish prefer the bottom, but adults are thought to spend more time actively swimming in the water column. The length at birth is 20 to 25 centimeters (8 inches); development is via aplacental viviparity (eggs held within the mother's body); litter size is up to sixty embryos; gestation is one year. The torpedo ray feeds on other fishes including sharks. The maximum size is about 180 centimeters (6 feet). Large individuals can produce a strong electric charge (220 volts) which could disable a sensitive person. Otherwise, they are relatively harmless.

DISTRIBUTION AND STATUS: They are distributed throughout the Gulf of Mexico but have probably been confused with the lesser electric ray. While it occurs in shallow water, it has been captured at over 500 meters (1,600 feet) depth. There is no information on population size and status.

SIMILAR SPECIES: The torpedo ray could be easily confused with the lesser electric ray. However, the short snout of the torpedo ray, almost straight margin of the disc ahead of the eyes, the large caudal, and the absence of dark blotches on the body separate it from the lesser electric ray.

ROUND STINGRAYS
FAMILY UROLOPHIDAE

The round stingrays are characterized by having a well-developed caudal fin, a round disc that is thick at its margin, a short thick tail, and, if present, a single dorsal fin. Only a single species, the yellow stingray, *Urobatis jamaicensis*, is found in the Gulf of Mexico.

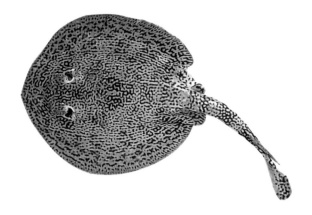

Yellow stingray, *Urobatis jamaicensi*

CHARACTERS: The dorsal surface of the yellow stingray is covered with fine reticulations of dark greenish, pinkish, or brown on a yellow or pale background. However, the color pattern varies widely even among individuals taken from the same area. This species has a somewhat oval disc that is slightly longer than it is wide and a relatively short, stout tail with a well-developed caudal fin. It has a moderately long snout.

BIOLOGY: The yellow stingray prefers shallow waters over sand and mud bottoms. My experience is that they seem to prefer reef areas. I have seen many around the reefs off of the Florida Keys and Belize, Central America. However, they are also found in estuaries and brackish areas. This fish grows to about 76 centimeters (2.5 feet). Males mature at about 30 centimeters (12 inches), and females most likely mature at a slightly larger size. Development is via aplacental viviparity, and three or four embryos have been observed. Yellow stingrays are known to feed on fishes, shrimps, mollusks, worms, and most likely, a variety of bottom-dwelling organisms. Feeding behavior may involve disturbing the sediment by undulating the disc and thus revealing prey items. The spine of the yellow stingray is apparently very toxic because fishers in the areas where these fish are found fear them. While these fish can inflict a very painful and serious wound, it is not normally life threatening.

DISTRIBUTION AND STATUS: The yellow stingray occurs out to about 25 meters (80 feet) depth. In the Gulf of Mexico it is common off of south Florida but is apparently absent in the northern and western Gulf. Only one species of Urolophidae is found in the Gulf of Mexico. There is no information available concerning status of this species.

SIMILAR SPECIES: The reticulated color pattern, round, thick disc, and short, thick tail separate this species from others in the area.

STINGRAYS
FAMILY DASYATIDAE

The stingrays are characterized by their very flattened angular disc, the absence of dorsal fins, and a long, thin tail that is armed with at least one spine (the "stinger") on its dorsal surface that is used as a defensive weapon. Most species have a single spine, but some have been found with as many as four. The spine has barbs directed rearward that prevent it from being removed easily. See page 46 regarding the method of treating stingray wounds and stingray safety. The stingrays are typically bottom-dwelling fishes, but some species do spend time actively swimming in the water column. There are seven Gulf of Mexico species in this family, but only six are considered here. The seventh species, *Himantura schmardae*, may be found in the southern Gulf, but its existence is not confirmed.

KEY TO THE SPECIES OF STINGRAYS

1. A. Leading edge of the disc is broadly rounded. Pelagic stingray, page 141.

A.

B. Leading edge of the disc is more angular forming a distinct point on the snout. Go to #2.

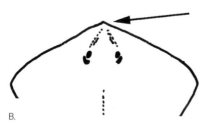

B.

2. A. Outer edges of disc are broadly rounded. Go to #3.

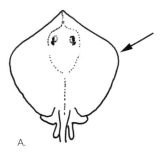

A.

B. Outer edges of disc are subangular, not rounded. Go to #4.

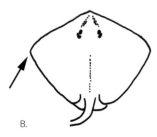

B.

3. A. Snout length (a) is longer than the length between the spiracles
(b). Atlantic stingray, page 139.
B. Snout length (a) is shorter than the length between the spiracles
(b). Bluntnose stingray, page 140.

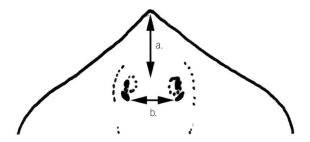

a.

b.

4. **A.** Many small, blunt spines (tubercles) found on middle area of disc; snout relatively long. Longnose stingray, page 138.

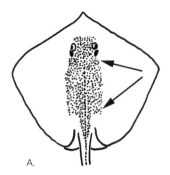

A.

B. Middle area of disc not covered with many small tubercles. Go to #5.

5. **A.** The tail is covered with many thorns, and the disc has a scattering of large thorns. Roughtail stingray, page 137.

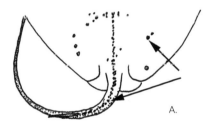

A.

B. Thorns found in a row extending along the midline of the disc from behind the eyes to the base of the tail. A short row of small thorns found about mid-disc on either side of the midline row. No thorns on side of tail. Southern stingray, page 136.

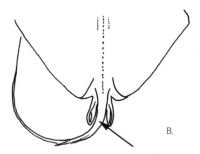

B.

Southern stingray, *Dasyatis americana*

CHARACTERS: The southern stingray varies from chocolate brown to olive green to gray above and white below. The disc is more angular than the other species in our area, having sharp outer corners. The tail is long and whiplike, and there is a single row of enlarged thorns that extend from just behind the eyes, along the midline, to the base of the tail. The tail also has a broad, black fold of skin along its base.

BIOLOGY: The southern stingray is a near-shore species that is frequently captured by hook-and-line fishers. It is a large ray, reaching a maximum of 150 centimeters (5 feet) disc width. Males mature at about 50 centimeters (1.6 feet) disc width, and females mature at about 80 centimeters (2.6 feet). Development is via aplacental viviparity; there are typically three to four embryos in a litter; and size at birth is about 18 centimeters (7 inches) disc width. It is found on sand, in seagrass meadows, in bays and estuaries, in lagoons, and on reefs. It feeds on bivalves (oysters, clams), worms, shrimps, crabs, and small fishes. It prefers to bury in the sand during the day and to forage at night. The large spine can inflict a serious wound.

DISTRIBUTION AND STATUS: This species is found out to about 100 m (330 feet) depth, and in the western Atlantic, from New Jersey, throughout the Gulf of Mexico, to Brazil. It is very common in much of the Gulf and is the ray that is encountered on the many "Swim with a Ray" excursions now offered in the Florida Keys and the Caribbean. The Gulf populations appear to be secure at this time.

SIMILAR SPECIES: This species could be confused with the roughtail stingray. However, the tail of the roughtail has many large thorns, the disc is not as angular in the roughtail, and the tail has a smaller ventral fold of skin. It might also be confused with the bluntnose stingray, although the outer corners of the disc are more evenly rounded in the bluntnose.

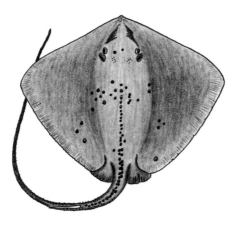

Roughtail stingray, *Dasyatis centroura*

CHARACTERS: The roughtail stingray varies from dark brown to olive brown above and white below. The disc is less angular than the southern stingray; the tail is long and whiplike; and there are many rows of thorns on the tail (hence the name "roughtail"). The tail also has a black fold of skin along its base, but this fold is much lower than on the southern.

BIOLOGY: The roughtail stingray is not common in the Gulf, but has been taken off Florida and Louisiana. It is a very large ray, reaching a maximum of 220 centimeters (7.2 feet) disc width. Males mature at about 140 centimeters (4.5 feet) disc width, and females mature at about 150 centimeters (5 feet). Development is via aplacental viviparity; there are typically two to six in a litter; and size at birth is about 35 centimeters (1 foot) disc width. It is found on sandy and muddy bottoms and is typically found near shore. It feeds on bottom-living organisms and small fishes. The large spine can inflict a serious wound.

DISTRIBUTION AND STATUS: This near-shore species is typically encountered out to about 100 meters (330 feet) depth, but it has been recorded to 200 meters (660 feet). Recorded from New Jersey, throughout the Gulf of Mexico, to Brazil. In the Gulf, this species is found in deeper water than other rays. The Gulf populations appear to be secure at this time.

SIMILAR SPECIES: See description of southern stingray.

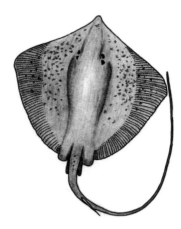

Longnose stingray, *Dasyatis guttata*

CHARACTERS: The longnose stingray is brown to yellowish or olivaceous, occasionally with dark spots above and whitish or yellowish below. This ray has a relatively long snout and a long, whiplike tail. The blunt tubercles that cover the middle region of the disc extending from around the eyes to the base of the tail are diagnostic. The ventral tail fold is black.

BIOLOGY: This is a poorly known species that apparently occurs only in the southern Gulf and perhaps in south Florida. It has a maximum size of about 200 centimeters (6.6 feet) disc width. Development is likely via aplacental viviparity, and a litter size of six embryos has been recorded. The large spine can inflict a serious wound.

DISTRIBUTION AND STATUS: In the western Atlantic this species is found in the southern Gulf of Mexico, perhaps southern Florida, and extends south to Brazil. It is most likely a shallow-water inhabitant. There is nothing known about population size or status.

SIMILAR SPECIES: This species could be confused only with the Atlantic stingray, but the band of blunt tubercles that cover the middle area of the disc between the eyes and the base of the tail readily separates it from that species.

Atlantic stingray, *Dasyatis sabina*

CHARACTERS: The Atlantic stingray is brown to yellowish above and white below. This stingray has a fairly long, triangular snout, and the corners of its disc are more broadly rounded. The Atlantic stingray's disc shape might be described as more "curvaceous" than other stingrays found in the Gulf. The tail is long and there is an upper and lower fold of skin on the tail. The tail spine(s) tend to be longer and more slender than related species. There is a row of enlarged blunt thorns along the back extending from the area behind the eyes out onto the tail.

BIOLOGY: This ray is one of the smallest in our area. It is born at about 10 centimeters (4 inches) disc width, and litter size is one to three young. Females mature at about 16 to 18 centimeters (6 to 7 inches) disc width. The largest individuals recorded were 46 to 61 centimeters (18 to 24 inches) disc width. These rays feed on worms, crabs, shrimps, and fishes. Their spine can cause a painful wound. Exercise caution when handling them.

DISTRIBUTION AND STATUS: In the western Atlantic this species is found from Chesapeake Bay to Brazil. This is a very common species in the shallow waters of the Gulf, inhabiting bays, estuaries, and lagoons and even entering fresh water. It is found from close to shore out to about 21 meters (70 feet). There is little information regarding Atlantic stingray populations, but they are very abundant in the Gulf.

SIMILAR SPECIES: The more rounded shape of the disc and its prominent snout separate it readily from the bluntnose stingray.

Bluntnose stingray, *Dasyatis say*

CHARACTERS: The bluntnose stingray has been described as yellowish, light brown, grayish, olivaceous, reddish brown, or dusky green. All of the specimens I have seen in the Gulf are brown. The underside is white. The bluntnose has a blunt, short snout and an angular disc. There is a well-developed fold on the upper and lower surface of the tail. There may be a row of small blunt thorns along the centerline of the back, but this changes with age. This is a moderate-sized ray.

BIOLOGY: The bluntnose stingray is a common ray in inshore waters, found in bays, estuaries, and along beaches, but also farther out to sea. It feeds upon bottom-dwelling organisms such as fish, clams, and shrimps. Maximum size is about 100 centimeters (3.3 feet) disc width. Litter size is two to four young, born at 15 to 16 centimeters (0.5 feet) disc width. It buries in the sand and, when trod upon by bathers, can inflict serious and painful injuries.

DISTRIBUTION AND STATUS: In the western Atlantic this stingray is occasionally recorded as far north as New Jersey throughout the Gulf of Mexico to Brazil, although it is not recorded from the coasts of Mexico, Central America, Colombia, or Venezuela. It is very common in the Gulf of Mexico, occurring from very shallow waters out to about 10 meters (33 feet). These rays are abundant in the Gulf, suggesting the species is not in peril.

SIMILAR SPECIES: This stingray could be confused with the Atlantic stingray. However, in the present species the margin of the disc from the snout to the tip of the pectorals is relatively straight whereas the margin in the Atlantic stingray is concave. See description of the southern stingray.

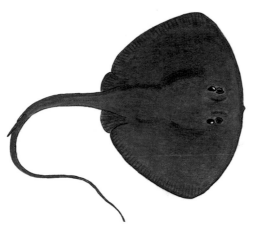

Pelagic stingray, *Pteroplatytrygon violacea*

CHARACTERS: The pelagic stingray has a blunt snout and a wide disc. The front margin of the disc forms an arch; the tail is thick at its base; the tail has no dorsal fold; the ventral tail fold is narrow; and there is a row of thorns along the midline of the disc from behind the eyes to the base of the tail spine. The pelagic stingray is purple, violet, or blue green upper and lower. There are no markings.

BIOLOGY: The pelagic stingray is not common in inshore waters in the Gulf of Mexico. It prefers the open ocean environment typically above 100 meters (328 feet) depth. This stingray does not typically lie on the bottom like others, but spends its time actively swimming. Prey include squid, fish, jellyfish, and crab. Maximum size is about 80 centimeters (2.6 feet) disc width. Females reach maturity at about 40 to 50 centimeters (1.3 to 1.6 feet) disc width and males at about 48 centimeters (1.6 feet) disc width. Litters consist of nine to thirteen young. Although this is a relatively harmless species, the large tail spine can inflict a painful wound.

DISTRIBUTION AND STATUS: In the western Atlantic this species is distributed from Grand Banks to Cape Hatteras, North Carolina, the northern Gulf of Mexico, and the Lesser Antilles. The status of this species is unknown.

SIMILAR SPECIES: The blunt snout and coloration of this species make it easily recognizable, and it should not be confused with other species in the area.

BUTTERFLY RAYS
FAMILY GYMNURIDAE

The butterfly rays have a very short tail that has rings of dark pigment around it, large expanded pectoral fins, and a very flattened body. They look sort of like a scale model of a U.S. Air Force "delta wing" airplane. A tail spine may or may not be present, depending upon species. Butterfly rays are found in tropical to warm-temperate latitudes in inshore waters including estuaries, bays, and river mouths. Although it is possible that the spiny butterfly ray, *Gymnura altavela*, is found in the Gulf, at present only a single species has been recorded.

Lesser butterfly ray, *Gymnura micrura*

CHARACTERS: The lesser butterfly ray has several distinctive characteristics: it has no tail spine, the front margin of its disc is somewhat concave, and the upper surface of its tail has a keel. The short tail has three or four dark crossbars. The color may be brown, gray, light green, or purple, with paler and/or darker vermiculations or dots above, and white below.

BIOLOGY: The lesser butterfly ray is not uncommon in inshore areas, inhabiting estuaries, bays, and off beaches. It is thought to prefer sandy-bottom habitat. Size at birth is about 20 centimeters (8 inches) disc width. Females are mature by 65 centimeters (25 inches) disc width (probably smaller), and males mature at about 42 centimeters (17 inches) disc width. Food items consist of fishes, crabs, shrimps, clams, and even items as small as copepods. This species is completely harmless.

DISTRIBUTION AND STATUS: In the western Atlantic this species has been recorded from North Carolina to the north coast of South America and is found throughout the Gulf of Mexico. This ray seems to be restricted to warmer, tropical/warm-temperate waters and apparently leaves the northern waters during cooler months. I have collected this little ray on a number of occasions along the northen Gulf of Mexico during summer. Nothing is known of its population size or status.

SIMILAR SPECIES: It is possible to confuse this species with the spiny butterfly ray, *Gymnura altavela*, although this species has not been recorded from the Gulf. However, the absence of a spine in the butterfly ray separates this species from all others.

SKATES
FAMILY RAJIDAE

The skates are often misidentified as rays, and no doubt many skates have met an untimely demise at the hands of fishers because of this confusion. However, the skates are completely harmless, lacking a stinging spine, and can be handled without fear. The skates look very much like the stingrays but skates possess a poorly developed caudal fin (rays have no caudal at all); they often have two dorsal fins located close to the end of the tail; they have no "stinging" tail spine; the tail is shorter and thicker than the rays; and there will frequently be thorns distributed over various regions of the body, particularly on the tail. Unlike the rays, the skates prefer saltier water and have not been reported from low salinity waters. All species of skates lay eggs. There are approximately twenty-two to twenty-four species in the Gulf of Mexico. However, because many of them prefer deep water, there are only about seven that might possibly be encountered and only two that are common in inshore waters. I have chosen to consider only those species that occur in water shallower than about 100 meters (330 feet).

KEY TO THE SPECIES OF SKATES

1. **A.** Tail long and slender, 65 to 70 percent of total body length. *Fenestraja sinusmexicanus*. Page 148.
   B. Tail short and relatively robust, 50 to 61 percent of total body length. Go to #2.

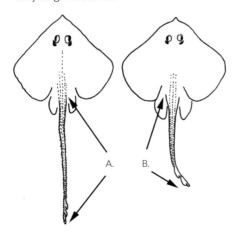

2. A. A single, ocellated (eyelike) spot on the upper surface of each
pectoral fin. Go to #3.
   B. No ocellated spots on pectoral fins. Go to #4.

3. A. Ocellated spots oval in shape. Ackley's skate. Page 151.
   B. Ocellated spots round in shape. Roundel skate. Page 153.

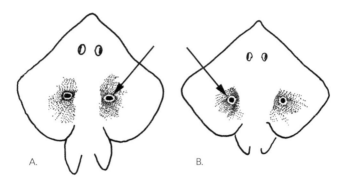

4. A. A single row of thorns along the midline of the back of the disc,
translucent area on either side of the snout. Clearnose skate.
Page 152.

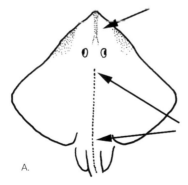

B. No or multiple rows of thorns along the midline of the back of
the disc. Go to #5.

5. **A.** No thorns along the midline of the back of the disc between the spiracles and the base of the tail (a single thorn may be present on the midline just behind the spiracles). Spreadfin skate. Page 147.

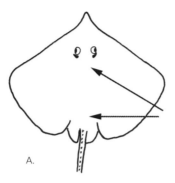

A.

B. Multiple rows of thorns along the midline of the upper surface of the disc. Go to #6.

6. **A.** Upper surface of disc with many conspicuous dark rosettes; tail with dark bars. Rosette skate. Page 149.
B. Upper surfaces of disc, tail, pelvics, and claspers covered with small, white, and light- to dark-brown freckles. Freckled skate. Page 150.

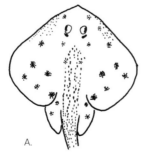

A.

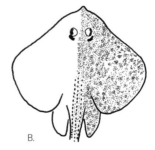

B.

Spreadfin skate, *Dipturus olseni*

CHARACTERS: The spreadfin skate is dark olive brown with small dark spots smaller than the pupil of the eye. The pectoral fins are broad; the snout is relatively long and slender, there are no thorns on the middle portion of the disc; the tail is relatively slender; and there are three rows of thorns on the tail.

BIOLOGY: There is virtually nothing known of this species' biology. Maximum size is about 57 centimeters (1.9 feet) total length, and males mature at about 50 centimeters (1.6 feet) total length. It is a harmless species.

DISTRIBUTION AND STATUS: In the western Atlantic this skate ranges from the northern Gulf of Mexico off the Florida panhandle to northern Mexico. It is a relatively deep-water species found between 55 and 384 meters (180 and 1,260 feet) depth. Nothing is known of its status.

SIMILAR SPECIES: This skate is easily distinguished from other species in that it is the only skate that has no thorns on the disc posterior to the spiracles.

Skate (No common name), *Fenestraja sinusmexicanus*

CHARACTERS: This skate is brownish purple, pale on either side of the snout, either plain or with small darker blotches. There may be one to four larger blotches on each side of the disc at about mid-disc position. The snout is short, and the tail is very long relative to body length. The tail has large thorns arranged in three rows. The tail is plain with no dark crossbars.

BIOLOGY: There is virtually nothing known of this species' biology. Maximum known size is about 36 centimeters (1.2 feet) total length. It is a harmless species.

DISTRIBUTION AND STATUS: In the western Atlantic, this skate ranges from the Gulf of Mexico, off Cuba, Nicaragua, and Venezuela. It occurs between about 59 and 1,096 meters (194 and 3,600 feet) depth. Nothing is known of its status.

SIMILAR SPECIES: The three rows of large thorns on the tail of this skate and the lack of dark crossbars on the tail separate it from other related species.

Rosette skate, *Leucoraja garmani*

CHARACTERS: The rosette skate is pale buff or brown with small light or dark spots and dark rosettes (flowerlike spots). The tail has dark crossbars. The snout is relatively short and blunt; the dorsal fins are similar; there are two to five rows of large thorns extending from about mid-disc along the tail to the origin of the first dorsal fin.

BIOLOGY: There is virtually nothing known of this species' biology. Maximum known size is about 34 centimeters (1.1 feet) total length. The rosette skate is a harmless species.

DISTRIBUTION AND STATUS: In the western Atlantic, this skate has been reported from Cape Cod to Nicaragua, including the Gulf of Mexico. It is a deep-water species found between 66 and 366 meters (217 and 1,200 feet) depth. Nothing is known of its status.

SIMILAR SPECIES: This skate is not easily confused with others because no skate in the western Atlantic has dark rosettes on the disc. The freckled skate is similar but is covered with many small spots (freckles) and does not have rosettes on the disc.

Freckled skate, *Leucoraja lentiginosa*

CHARACTERS: The upper surface of the freckled skate is covered with many small light- to dark-brown freckles and whitish spots. The tail, pelvics, and claspers are likewise spotted. The tail has dark crossbars. The lower surface is whitish. The snout is relatively short and blunt; the dorsal fins are similar; there are three to five rows of large thorns extending from about mid-disc along the tail to the origin of the first dorsal fin.

BIOLOGY: There is virtually nothing known of this species' biology. Maximum known size is about 42 centimeters (1.4 feet) total length, and maturation occurs at about 35 to 42 centimeters (1.1 to 1.4 feet). The freckled skate is a harmless species.

DISTRIBUTION AND STATUS: In the western Atlantic, this skate is apparently known only from the Gulf of Mexico, where it is widespread. It is a deep-water species found between 53 and 588 meters (174 and 1,930 feet) depth. Nothing is known of its status.

SIMILAR SPECIES: See description of the rosette skate.

Ackley's skate, *Raja ackleyi*

CHARACTERS: Ackley's skate is pale yellowish brown with small light and dark spots. There is a dark, oval, ocellated (looks like an eye) spot with a pale margin on each pectoral. There is a cluster of small thorns on the tip of the snout and a midline row of thorns extending from behind the eyes to the first dorsal fin. There may be one or two rows of thorns on either side of the tail.

BIOLOGY: There is virtually nothing known of this species' biology. A mature male of 41 centimeters (16 inches) total length has been recorded. It is very uncommon in the Gulf of Mexico. It is harmless.

DISTRIBUTION AND STATUS: This skate has been reported from the Yucatan, in southern Florida, and off the Dry Tortugas. It is a deeper-water species found between 32 and 384 meters (100 and 1,260 feet) depth. Nothing is known of its status.

SIMILAR SPECIES: This skate could be confused with the roundel skate. However, the two prominent oval spots on the upper surface of Ackley's skate easily separate it from the roundel skate, which has round spots.

Clearnosed skate, *Raja eglanteria*

CHARACTERS: The clearnosed skate has a translucent space on either side of the snout, which gives this species its name. The snout is elongated and pointed. The tail is relatively thick and has narrow lateral folds along its entire length. The two dorsal fins are nearly equal in size and are located near the end of the tail. The caudal is poorly developed. There is a single row of thorns along the midline of the back. It is brown above with many dark brown spots, elongate bars, and light brown irregular spots. The tail has wide, brown crossbars.

BIOLOGY: This is the most common skate in the shallow waters of the Gulf of Mexico. However, they are rarely taken in bays or estuaries because they prefer waters of higher salinity. This skate feeds on clams, crabs, shrimps, worms, squids, and fishes. The egg cases are dark in color and somewhat oblong in shape, and they have long, stiff tendrils at each corner. Maximum size is about 79 centimeters (2.6 feet); females mature at 60 to 78 centimeters (2.0 to 2.6 feet), and males mature at 54 to 77 centimeters (1.8 to 2.5 feet). Young hatch at about 14 centimeters (5.5 inches). The clearnosed skate is harmless to humans.

DISTRIBUTION AND STATUS: In the western Atlantic, this skate is found from Massachusetts to Florida and the northern Gulf of Mexico. It is more common east of the Mississippi River. It is found from the shore to 119 meters (390 feet) depth. This species is likely secure in the Gulf of Mexico.

SIMILAR SPECIES: The clearnosed skate could not be easily confused with the common skate species in the Gulf of Mexico.

Roundel skate, *Raja texana*

CHARACTERS: The roundel skate has a moderately long snout; a disc that has abruptly rounded outer corners; a narrow, lateral tail fold; dorsal fins of similar size and shape, and a row of thorns that extends along the midline from just behind the eyes to the first dorsal fin origin. The upper surface is brown with a round ocellus (eyelike spot) on each pectoral fin; the ocellus is brown or black surrounded by a yellow ring. The underside is white.

BIOLOGY: This skate is a relatively common resident of the Gulf of Mexico. It is not found in lower salinity waters so it is not likely to be found in estuaries in the Gulf. It feeds on crustaceans, fishes, and various bottom-dwelling organisms. Young may be found in bays, but adults move offshore in deeper waters. The egg capsules are dark in color, somewhat oblong in shape, with filaments at the corners. Maximum size is about 54 centimeters (1.8 feet) in length. Males mature at about 47 centimeters (1.5 feet) in length.

DISTRIBUTION AND STATUS: The roundel skate is found from shallow waters to about 180 meters (590 feet) depth, but it is more prevalent in waters less than 90 meters (295 feet) depth. In the western Atlantic it is found off the west coast of Florida and throughout the Gulf of Mexico to the Yucatan bank. The species does not appear to be in any danger of extinction, although little information is available.

SIMILAR SPECIES: See the description of Ackley's skate.

EAGLE RAYS
FAMILY MYLIOBATIDAE

Unlike the stingrays and skates that prefer the bottom of the ocean, the eagle rays are adapted to spend more time swimming. They have large pointed wings that are wider than the length of the body and a long, slender tail. There is a small dorsal fin near the base of the tail. The head is distinctly elevated above the disc, and the eyes are located on the side of the head. The tail has no caudal fin. To see these rays swim reminds one of bird flight because they seem to fly through the water by flapping their large pectoral fins. There are twenty-four species of eagle rays worldwide, but only two are found in the Gulf of Mexico. The Gulf species have live birth.

KEY TO THE SPECIES OF EAGLE RAYS

1. **A.** Disc black, bluish, or brown and covered with many small white, greenish, or yellow spots. Teeth in a single series in the jaw. Spotted eagle ray, page 155.
   B. Disc grayish, reddish brown, or dark brown with diffuse whitish spots. The posterior part of the tail is dark brown or black. The teeth are green. Bullnose ray, page 156.

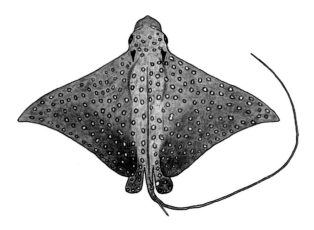

Spotted eagle ray, *Actobatis narinari*

CHARACTERS: The spotted eagle ray has large, expanded pectoral fins with pointed tips. The head and snout are distinct from the body; the head is elevated above the disc; and the snout is thin and elongate like a duck's bill. There is a long, slender tail with a long spine near the base. The color is black, bluish, brown, or olivaceous with many small white, greenish, or yellow spots.

BIOLOGY: The spotted eagle ray is commonly found in bays, in estuaries, and around coral reefs. I have seen these rays cruising the grass flats in the Florida Keys on many occasions but have never seen one in the northern Gulf of Mexico. They are often seen swimming close to the water's surface and are known for leaping out of the water. They feed on shrimps, clams, oysters, snails, and fishes. Litter size is typically four, and young are born at about 17 to 35 centimeters (0.6 to 1.1 feet) disc width. Maximum size is 280 centimeters (9.2 feet) disc width. The spine can inflict a serious wound, but otherwise they are harmless.

DISTRIBUTION AND STATUS: This ray is distributed in the western Atlantic from North Carolina to southern Brazil. It is not common in the northern Gulf. Its depth range is from near shore to about 80 meters (260 feet). The spotted eagle ray has likely experienced reductions in numbers, but little information is available. It is listed as threatened by some international wildlife agencies.

SIMILAR SPECIES: The distinctive coloration of this species makes it difficult to confuse with any other. It is possible that the bullnose ray may be confused with this species, but the coloration and the green teeth of the bullnose are distinctive. Also, there is a single series of teeth in the jaws in the spotted eagle ray whereas the bullnose ray has more than one series of teeth.

Bullnose ray, *Myliobatis freminvillei*

CHARACTERS: The bullnose ray has large, expanded pectoral fins with pointed tips. The head and snout are distinct from the body; the head is elevated above the disc; and the snout is thin and elongate like a duck's bill. There is a long, slender tail with a long spine near the base. The color is grayish, reddish brown, or dark brown with diffuse whitish spots. The posterior part of the tail is dark brown or black. The teeth are green.

BIOLOGY: The bullnose ray is commonly found in coastal waters. It has been known to leap from the water. The bullnose feeds on shrimps, lobsters, crabs, clams, and oysters. Litters of six young have been reported, and young are born at about 25 centimeters (0.8 feet) disc width. Maximum size is about 86 centimeters (10 feet) disc width. The spine can inflict a serious wound, but otherwise they are harmless.

DISTRIBUTION AND STATUS: This ray is distributed in the western Atlantic from Cape Cod to Brazil. In the northern Gulf it appears to be more common east of the Mississippi River, although it is not particularly common in any part of the Gulf. Its depth range is from near shore to about 10 meters (33 feet) depth. This ray has likely experienced reductions in numbers, but little information is available.

SIMILAR SPECIES: See description of the spotted eagle ray.

COWNOSED RAYS
FAMILY RHINOPTERIDAE

In the Gulf of Mexico, there is only a single species in this family, the cownosed ray, although worldwide there are about five species. All cownosed rays have a greatly expanded disc that is wider than it is long; the head is higher than the disc; and the eyes are located on the side of the head. There is a small dorsal located near the base of the tail. The tail spines may be multiple and are serrated.

Cownosed ray, *Rhinoptera bonasus*

CHARACTERS: This is a characteristic ray possessing a deep groove around the front of the head below the eyes, and the snout below the groove has two lobes. The wings are long and pointed. The tail is very long and slender with one or more spines at the base of the tail close to the body. The color is uniformly brown to olive brown on the upper surface and white or yellowish white below.

BIOLOGY: This is a very common species in the Gulf of Mexico. I have seen huge schools of cownose rays off of Florida, Alabama, and Mississippi. These rays migrate long distances in the fall from Florida to Yucatan, Mexico. They occasionally make dramatic leaps from the water. Food items include crustaceans, clams, and snails. Maximum size is about 86 centimeters (2.8 feet) disc width; males mature at 60 centimeters (2 feet) disc width. These fish have live birth, and young are born at about 25 centimeters (10 inches) disc width. In the northern Gulf newborn specimens appear in the late spring and early summer. When captured, cownosed rays make a sound sort of like a duck quacking, and they also excrete an incredible amount of viscous mucous. The tail spine could inflict a painful wound, but otherwise it is harmless.

DISTRIBUTION AND STATUS: In the western Atlantic, the cownosed ray is found from southern New England to southern Brazil including the Gulf of Mexico. Depth distribution is from near shore to 22 meters (72 feet). This ray is likely in no immediate danger of extinction.

SIMILAR SPECIES: The cownosed ray is an unmistakable species not easily confused with any other. The eagle rays do not have the deep groove around the head, and the mantas have the characteristic "horns" on the head.

MANTA AND DEVIL RAYS
FAMILY MOBULIDAE

The manta and devil rays are perhaps the most widely known rays because of their extremely large size and unique appearance. The cephalic fins that extend forward from the head give these rays an evil appearance, hence the name "devil rays." The head is short, broad, and slightly raised above the body. The tail is short and slender with a small dorsal fin near its base. The disc is wider than it is long, and the pectorals are pointed. There is no caudal fin. The mantas have live birth. They are found in all the world's oceans in tropical to warm-temperate waters. There are about thirteen species, but only two are recorded from the Gulf of Mexico.

KEY TO THE SPECIES OF THE MANTA AND DEVIL RAYS

1. **A.** Deeply concave region between the large cephalic fins (horns), head very broad. Manta ray, page 160.
   B. Slightly concave region between the cephalic fins (horns); head relatively narrow. Devil ray, page 161

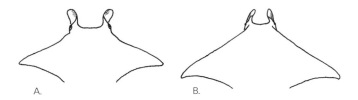

A.   B.

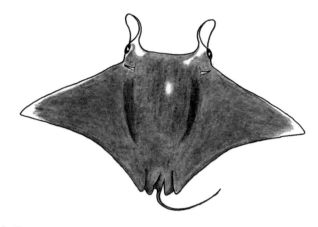

Manta ray, *Manta birostris*

CHARACTERS: The manta is the world's largest ray. It is readily recognized by its very large head; large, hornlike cephalic fins; a deeply concave head between the cephalic fins; a terminal mouth; a very short, slender tail usually with no spine; a disc wider than it is long; and pointed pectorals. The color is blackish, reddish, or brownish above occasionally with white shoulder patches. The ventral surface is white occasionally with gray blotches.

BIOLOGY: This species is typically seen in inshore waters around coral and rocky reefs but may be found over very deep water and apparently in shallow muddy bays and off river mouths. This is the ray that is large and approachable enough for skin divers to actually hitch rides on them. The manta ray swims in somersaults with its mouth open and filters plankton (small plants and organisms) from the water for food but may also consume fishes. It makes spectacular leaps from the water and has been known to leap into open fishing boats with disastrous results. Maximum size is about 800 centimeters (26 feet) disc width. Young are born at 110 to 130 centimeters (3.6 to 4.3 feet) disc width. Litter sizes are up to two young. Other than the situation above, this species is harmless to humans.

DISTRIBUTION AND STATUS: In the western Atlantic, the manta ray is found from southern New England to Brazil. It is not commonly seen in the shallow inshore waters of the Gulf of Mexico.

SIMILAR SPECIES: This species could be confused with the devil ray, but the devil ray has a much narrower head and the region between the cephalic fins is only slightly concave.

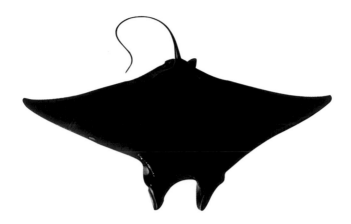

Devil ray, *Mobula hypostoma*

CHARACTERS: The devil ray has a moderately expanded head, relatively small cephalic fins (the hornlike extensions on the head), and a very slender and moderately long tail with no spine. The disc is wider than it is long, and the pectorals are pointed. The color is very dark purple to black above and whitish below.

BIOLOGY: This is a very uncommon species in the Gulf of Mexico. In many years of collecting I have captured only one specimen off the south beach of Dauphin Island, Alabama. This ray is an inshore inhabitant, but it is perhaps more common over deeper waters. It feeds by filtering small organisms from the water but may also take small fish. When captured, this ray makes an interesting sound somewhat like a single note from a brass bell. Maximum size is about 120 centimeters (4 feet) disc width. A female with a single embryo has been recorded. This species is harmless to humans.

DISTRIBUTION AND STATUS: In the western Atlantic this ray occurs from New Jersey to Brazil including the Gulf of Mexico. The status of this species is unknown.

SIMILAR SPECIES: See description for manta ray.

PHOTO CREDITS

Illustrations courtesy of the author unless otherwise noted

Page ii: Caribbean reef shark, courtesy of John D. Hewitt. Page 26: Cuban dogfish, courtesy of Lew Bullock. Page 60: blacknose shark, courtesy of Jill Frank. Page 61: bignose shark, courtesy of Eric Hoffmayer. Page 69: dusky shark, courtesy of Eric Hoffmayer. Page 72: smalltail shark, courtesy of Steve Branstetter. Page 86: smooth hammerhead shark, courtesy of Steve Branstetter. Page 88: seven-gill shark, courtesy of Michael Hendon, NOAA, National Marine Fisheries Service Pascagoula, Mississippi, Laboratory. Page 92: angel shark, courtesy of Jim Bartlett. Page 94: common thresher, courtesy of Eric Hoffmayer. Page 99: shortfin mako, courtesy of Steve Branstetter. Page 101: longfin mako, courtesy of Steve Branstetter. Page 105: sand tiger shark, courtesy of Eric Hoffmayer. Page 106: bigeye sand tiger shark, courtesy of Steve Branstetter. Page 110: Florida smoothhound shark, courtesy of Eric Hoffmayer. Page 111: smooth dogfish, courtesy of Eric Hoffmayer. Page 130: Atlantic torpedo, courtesy of Michael Hendon, NOAA, National Marine Fisheries Service Pascagoula, Mississippi, Laboratory. Page 149: Rosette skate, courtesy of Michael Hendon, NOAA, National Marine Fisheries Service Pascagoula, Mississippi, Laboratory. Page 150: freckled skate, courtesy of Michael Hendon, NOAA, National Marine Fisheries Service Pascagoula, Mississippi, Laboratory. Page 152: clearnosed skate, courtesy of Jim Bartlett. Page 153: roundel skate, courtesy of Michael Hendon, NOAA, National Marine Fisheries Service Pascagoula, Mississippi, Laboratory.

INDEX